Energy, Environment and Sustainable Development

About the Author

Pradeep Chaturvedi, the Chairman of The Institution of Engineers (India), Delhi State Centre is a Mechanical Engineer and an Energy and Environment Expert. Mr. Chaturvedi has been involved with sustainable development issues, mainly the energy policy and planning and implementation of environment friendly sustainable energy projects for over three decades in India and other Asian and Pacific countries. As a Member of the Perspective Planning Group of Planning Commission, preparing Development Vision 2020, he contributed 'India's Energy Vision 2020'. He has been actively involved with the formulation of Energy Conservation Act 2001 and the Electricity Bill 2001.

Mr. Chaturvedi has been involved with the Studies Programme of the World Energy Council since 1983. His understanding of energy issues of the South Asia Region, in the global context, has been widely appreciated.

Mr. Chaturvedi was deeply involved with the preparatory phase of the World Summit of Sustainable Development. He is also involved in the follow-up phase.

Mr. Chaturvedi is Vice-Chairman of the UNESCO sponsored World Renewable Energy Network; and Vice-Chairman of the World Environment Foundation, both based in UK; and President of the Bio Energy Society of India. He is also a Visiting Faculty at IIT, Delhi.

Mr. Chaturvedi has authored six and edited 25 books on Energy Management and Sustainable Development. These books are widely referred to understand the situation in the country.

Energy, Environment and Sustainable Development

PRADEEP CHATURVEDI

CONCEPT PUBLISHING COMPANY, NEW DELHI-110059

ISBN 81-8069-033-4

First Published 2003

Published and Printed by
Ashok Kumar Mittal
Concept Publishing Company
A/15-16, Commercial Block, Mohan Garden
New Delhi-110059 (INDIA)

Phones : 25351460, 25351794
Fax : 091-11-25357103
Email : publishing@conceptpub.com

Preface

Energy services are essential for sustainable development. The way these services are produced, distributed and used affect the social, economic and environmental dimensions of development. Although, energy itself is not a basic human need, it is critical for fulfillment of all needs. Lack or access to diverse and affordable energy services means that the basic needs of the people are not being met.

Energy is a vital entry point for the improvement of quality of life of the poor. It is also strongly linked with economic growth, the pillar of sustainable development. Energy services are needed to create jobs, develop industries, enhance value added economic activities and support income earning activities in rural areas. Fuels are essential for heat using processes, transportation and many industrial activities. Electricity is an essential input to modern productive services as well as communication and service industries.

Lack of energy supply and interruptions of energy supply cause serious, financial, economic and social losses. Energy must be available at all times, in sufficient quantities and at affordable prices, to support the goals of sustainable development. From a balance of payments perspective, energy imports are currently one of the largest sources of foreign debt for many of the poorest countries. In addition, investments that were made in high cost centralized conventional energy installations have contributed to the growth of foreign debt.

Attention to energy security is critical because of the uneven distribution of both, fossil fuel resources and the capacity to develop other resources. Whereas energy efficiency improvements have the scope for cost effective energy availability, it has the limitation to its contribution.

Realising the increasing need of our economic and social sectors, the Government of India, in the early 90s decided to bring in the reforms in the energy sector. The process of reforms was introduced at a rapid pace in the power sector, mainly for capacity addition. However, it was realized that the main thrust of the reforms being to attract private sector and foreign investment was not moving at the desired pace. The

Government therefore shifted its thrust from power generation to power transmission and distribution. The second phase of the reforms process has started only in the year 2001 and their impact is still to be seen.

The reforms process was really launched when Agenda-21 had shown concern for climate change and desired various countries to act sensibly and in an environmentally friendly manner. Thus, the concept of sustainable energy supply took roots in the second half of the 90s. Present developments in the energy sector are linked to sustainability issues reflected also at the World Summit on Social Development.

Coinciding with the reforms process and developments in environment sector the author had the privilege to address on different platforms covering the whole canvas of reforms and energy-environment interactions. This Book is a compendium of all such papers, mainly presented during the last two years.

The book also includes a Platinum Jubilee Lecture by Mr. Gerald Doucet, Secretary General, World Energy Council, on the topic 'Energy for People, Energy for Peace—Markets Sustainability and Drivers', delivered at the Institution of Engineers (India), Delhi State Centre. In view of the understanding of various issues by the World Energy Council, the Paper presents valuable additional information, which has been added as an annexure to the book.

I hope this Book containing papers presented by the author as Chairman of The Institution of Engineers (India), Delhi State Centre forms an important part of the Platinum Jubilee Celebration of the Centre.

PRADEEP CHATURVEDI
Chairman
The Institution of Engineers (India)
Delhi State Centre, New Delhi

New Delhi
December 23, 2002

Acknowledgements

I gratefully acknowledge the help and support of my colleagues at the Institution of Engineers (India), Delhi State Centre for motivating me to write this book and have it released on the occasion of the Annual Function on December 23, 2002. My special thanks to Shri V.K.Gupta, the in-coming Honorary Secretary of the Delhi State Centre for his persuasion to complete and publish the book.

Help and support of Shri Subhash Chander, Honorary Secretary, Delhi State Centre and Shri Praveen Singh, the Officer Administration, Delhi State Centre at different stages of preparation of papers and presentation is duly recognized. Secretarial assistance of Shri U.K. Bhalla is also duly recognized.

Shri Ashok K. Mittal, Managing Director, Concept Publishing Company deserves special thanks for his effort to publish this book at a very short notice.

My thanks also to various Institutions and personalities, who invited me to make these presentations during my tenure as the Chairman, The Institution of Engineers (India), Delhi State Centre.

PRADEEP CHATURVEDI
Chairman
The Institution of Engineers (India)
Delhi State Centre, New Delhi

New Delhi
December 23, 2002

Contents

PART I : ENERGY

PART II: ENVIRONMENT

PART I

ENERGY

1

Issues in Sustainable Energy Supply in India

Introduction

India launched its energy sector reforms programme in 1992 with a thrust on power sector reforms. A major initiative was launched on September 15, 2000 — The Engineers Day — when the largest investment commitment was signed between Reliance Power Ltd.0 of India and the Southern Energy Incorporate of USA to develop the prestigious Hilma Power Project in Orissa (in eastern part of India) with a installed capacity of 3,690 MW. The State owned Power Trading Corporation will sell power from Hilma Power Project to federal states of Punjab, Haryana, Madhya Pradesh, Gujarat and Rajasthan.

Another major agreement was signed on the same day between Power Trading Corporation of India and CMS Energy Corporation of USA who seeks to sell the entire power from the Tamil Nadu power plant of capacity of 1,850 MW. The CMS Energy Corporation will also develop liquefied natural gas terminal at Ennore in the State of Tamil Nadu.

These two major power projects can be a turning point in India's efforts to privatise the power sector in the country. After 8 years of intensive efforts, the total power generation plants commissioned by private sector (till September 1, 2000) were 5058 MW. Another 18 projects with a total capacity of 5214 MW are in various stages of construction. Several other projects are awaiting statutory clearance.

Unnecessary focus on Dhabol Power Project (Enron Project) has overshadowed all such initiatives. Too much publicity of Dhabol Power Project tends to give an impression that the Indians are the culprits and if that project stops or is sold off than that is the end of

---

Paper presented at the All India Seminar on 'Power Development in India' at the 10th Annual General Conference of the Institution of Engineers (India), Agartala Local Centre on August 10-12, 2001, Agartala.

foreign investments in India. Neither was the Dhabol Power Project the first FDI nor will it be the last FDI. Should our legal and administrative regime improve alongwith the discipline and maturity of the capital markets there should be no problem in attracting FDIs.

Another important point at this stage. All the major power producing companies in the Public Sector today have improved their operational and financial status to the stage that the financial institutions are ready to offer the investments required by them. In my personal discussion with several of these companies including NHPC, NTPC, BSES, Powergrid I have been informed in no uncertain terms that *financing in the power sector in no longer a critical issue. No doubt it remains the crucial issue.*

The confidence of all such utilities is a welcome sign. If government so desires they can highlight it. However it remains crucial for the Seminar on 'Power Development in India' being organised on the eve of the Independence Day. Neither privatisation is all rosy nor is public sector all inefficient. The Seminar needs to focus on the central issue of financing of power projects and speedy clearances for the project proposals.

The energy development has attracted large public funds in the planned growth of the economy. Efforts to promote use of fossil fuel (in an efficient fashion) and of new and renewable sources of energy have ensured a maximum supply of energy in the country. The decade of 90s has witnessed restructuring of the energy sector and gradual entry of private sector companies. The transition is accompanied by teething troubles and adjustment to changing circumstances.

Commercial Energy Resources

Primary Energy Resources and Regional Distribution

Coal and hydroelectric resources are the mainstay for India's primary resource support for secondary energy purposes. Oil and natural gas resources of the country are limited. Nuclear resources, on the other hand, are modest. The coal resources are of poor quality having high ash content. Besides, the distribution of the primary resources in the various regions is rather skewed. While the eastern region accounts for 70 per cent of the total coal resources, the western region accounts for over 70 per cent of the hydrocarbon reserves, and over 70 per cent of the total hydro potential of the country is concentrated in the northern and north-eastern region. The southern region which has only 6 per cent of the coal reserves and 10 per cent of the hydroelectric potential, possesses most of the lignite deposits of the country.

Refer to Table 1.1 for regional distribution of the primary commercial energy.

Table 1.1: Primary Commercial Energy Resources (March 1997)

Regional	*Coal (BMT)*	*Lignite (BMT)*	*Crude Oil (MMT)*	*Natural Gas (BCM)*	*Hydro Electricity*	
					Capacity (x) MW	*Energy TWh*
Northern	1.1	1.1	—	4	30135	225.00
Western	48.2	0.5	584	497	5679	31.40
Southern	13.1	25.9	—	—	10763	61.80
Eastern	141.4	—	—	—	5590	42.50
North-Eastern	0.9	—	148	159	31857	239.30
All India	204.7	27.5	732	660	84044	600.00

Source: Planning Commission, Government of India (1998); and ONGC (1999)
BMT = Billion Metric Tonnes MMT = Million Metric Tonnes;
BCM = Billion Cubic Metre (x) at 60 per cent LF.
Note: Gas reserves are now identified at 717.3 BCM (at the end of 1998)— Under control of: ONGC — 525 BCM; OIL — 94.5 BCM; & Joint Ventures/Private companies — 97.5 BCM.

Non-conventional Energy Sources

The country is endowed with a large potential of non-conventional energy resources are given in Table 1.2.

Trends in Energy Production, Supply and Consumption

Indigenous Primary Energy Production

India's energy requirements are met from traditional as well as modern forms of energy. The share of traditional forms of fuels is diminishing, and in 1996-97 these contributed to over 34 per cent of the annual energy consumption. The usage is largely restricted to domestic fuels in semi-urban and rural areas at very low efficiencies. Current trends in production are given in Table 1.3.

Table 1.2: Renewable Energy Potential and Exploitation (in March 1999)

Source/Technology	*Potential Availability*	*Potential Exploited*
Biogas Plants	12 Million	2.713 Million
Biomass-based Power	17,000 MW	114 MW
Efficient Woodstoves	120 Million	2.8.5 Million
Solar Energy	5 × 10 Whr/Year	32 MW SPV 4,20,000 Sq Mtrs SWH 4,57,000SC
Small Hydro	10,000 MW	155 MW
Wind Energy	20,000 MW	970 MW
Energy Recovery from Urban & Industrial Wastes	—	4.75 MW
Ocean Thermal	50,000 MW	—
Sea Wave Power	20,000 MW	—
Tidal Power	9,000 MW	—

Source: Ministry of Non-Conventional Energy Sources, Government of India. Annual Report 1998-99.

SPV = Solar Photovoltaic; SWH = Solar Water Heater; SC = Solar Cooker

Table 1.3: Trends in Indigenous Production of Commercial Primary Energy

Year	*Coal Mt*	*Lignite Mt*	*Crude oil mt*	*Natural Gas mt*	*Hydro TWh*	*Nuclear TWh*	*Wind TWh*
1953-54	33.0	—	0.26	—	2.52	—	—
1960-61	55.67	0.05	0.45	—	7.84	—	—
1970-71	72.95	3.39	6.82	1.4	25.25	2.42	—
1980-81	114.1	4.80	10.51	2.4	46.54	3.00	—
1990-91	211.73	14.07	33.02	18.0	71.66	6.14	0.03
1996-97	288.65	22.54	33.87	22.9	68.63	9.01	0.85
1997-98	295.9	23.05	33.86	26.40	74.5	10.00	1.00
1998-99	292.2	23.4	32.7	27.43	82.7	12.00	1.00

Source: Planning Commission, Government of India (1998); Indian Economic Survey 1999-2000 (2000); and ONGC (1999).

Total power generation capacity installed on March 31, 1999 was 93,300 MW as indicated in Table 1.4. These plants generated a total of 448.4 bkwh of electricity in 1998-99 comprising of thermal — 353.7 bkwh; hydro — 82.7 bkwh and nuclear — 12.0 bkwh. Thermal plants account for 79.9 per cent; hydro electric plants for 17.7 per cent and nuclear plants for 2.4 per cent of power generation. This recorded a growth of 6.6 per cent over the previous year.

Table 1.4: Installed Plant Capacities and Electricity Generation (March 1999)

Source	*Installed Capacity*	*Generation*
Utilities		
Hydro	22,400 MW	82.7 Bkwh
Thermal	68,700 MW	353.7 Bkwh
Nuclear	2,200 MW	12.0 Bkwh
Sub-Total	93,300 MW	448.4 Bkwh
Non-Utilities	14,500 MW	46.0 Bkwh
Total	107,800 MW	494.4 Bkwh

Source: Indian Economic Survey 1999-2000 (2000).

Note: Figures may not add up to the total owing to rounding off.

Energy Imports

Energy imports are largely in the form of crude oil and oil products. The import/export of other commercial energy forms, viz. coal and secondary electricity, has been rather limited and restricted.

The import of crude oil and oil products in the year 1997-98 has been 34.49 mt and 19.5 mt (crude equivalent) respectively. Import of coal during 1997-98 was 17 mt.

Primary Energy Supply

Total supply of energy (both of commercial, and non-commercial forms excluding draught animal power) increased from 89.6 mtoe in 1953-54 to about 377 mtoe in 1996-97. Share of non-commercial fuels declined from 74 per cent in 1950-51 to about 34 per cent in 1996-97. Fuelwood continues to maintain its primacy in the non-commercial energy supply (largely in the household sector). (The break-up for 1997-98 has still to be worked out as the figures for energy production and supply are still provisional.) Changes in the pattern of primary energy supply since India's planned development in 1952 is reflected in Table 1.5.

Table 1.5: Changes in the Pattern of Primary Energy Supply (mtoe)

Commercial Primary Energy Production	*1953-54*	*1960-61*	*1970-71*	*1980-81*	*1990-91*	*1996-97*	*1997-98 (x)*
Coal	23.62	35.64	36.48	56.96	94.68	122.79	125.79
Lignite	—	0.01	0.81	1.23	3.34	5.04	5.20
Crude Oil	0.19	0.46	7.01	10.79	33.92	33.87	33.86
Natural Gas	—	—	0.60	1.41	11.73	20.05	24.70
Hydro Power	0.24	0.67	2.17	4.00	6.16	5.90	6.45
Nuclear Power	—	—	0.63	0.78	1.60	2.35	2.50
Wind Power	—	—	—	—	—	0.07	0.10
Total	**24.05**	**36.78**	**47.67**	**75.19**	**151.43**	**190.30**	**198.60**
Net Imports	2.20	6.04	12.66	24.63	31.69	61.63	61.63
St. Changes	(-) 0.24	2.87	0.69	3.80	5.37	2.38	2.38
Intl. Bunkers	(-) 0.53	2.50	0.24	0.21	0.14	0.17	0.17
Total Commercial Energy Supply	**25.48**	**39.45**	**59.40**	**95.81**	**177.61**	**249.38**	**262.7**
Non-Commercial Primary Energy Supply	**64.13**	**74.38**	**86.72**	**108.48**	**122.07**	**127.51**	**127.5**
Total Primary Energy Supply	**89.61**	**103.83**	**146.12**	**204.29**	**299.68**	**376.89**	**390.12**

Source: Planning Commission, Government of India (1998). (x) Estimated/Provisional.

Note: Supply (also considered consumption) of non-commercial fuels in the year 1996-97 is estimated at 173 mt of fuelwood, 95 mt of dung cakes and 53 mt of agro waste.

Energy Consumption

Share of commercial energy in the final energy consumption increased from 28.4 per cent to 65.9 per cent during this period, while that of non-commercial energy sources has declined from 71.6 per cent (in 1953-54) to 34.1 per cent (in 1996-97) (see Table 1.6).

Table 1.6: Changes in Pattern of Energy Consumption (in per cent)

Year	*Main Energy Form*		*Break-up of Commercial Energy Consumption*				
	Non-Comml.	*Comml.*	*Coal*	*Oil*	*Natural Gas*	*Electricity*	*Total*
1953-54	71.6	28.4	80.1	16.7	0.0	3.2	100.0
1960-61	65.3	34.7	75.3	19.9	0.0	4.8	100.0
1970-71	59.4	40.6	56.1	34.1	0.6	9.2	100.0
1980-81	53.1	46.9	47.9	40.3	1.1	10.7	100.0
1990-91	40.7	59.3	35.9	43.6	5.5	15.0	100.0
1996-97	34.1	65.9	29.3	46.8	6.9	17.0	100.0

Source: Planning Commission, Government of India (1998).

The total consumption of petroleum and petroleum products during 1997-98 is reported at 84.5 mt (80.1 mt from the public sector units and 4.4 mt imported by the private parties).

Energy Policy

Strategic Elements of Energy Policy

Energy strategies adopted, from time to time are preceded by detailed energy studies by expert groups. The first such study was carried out in 1961, under aegis of the *Energy Survey of India Committee*, with a comprehensive survey of energy resources of the country which enabled formulation of a pragmatic energy strategy during the sixties. During the period between the two oil shocks (1973-79) the energy situation was reviewed twice — once under the aegis of the Fuel Policy Committee, in 1974, and again under the aegis of the Energy Policy Committee in 1979. In 1976, a comprehensive Energy Policy was formulated.

The Ninth Five Year Plan (1997-2002) has defined that the energy strategy for the future could be divided into short-term strategy, medium-term strategy and long-term strategy.

Various components of short/medium-term strategy could be:

- The tariff structure of various energy products to be rationalised, particularly the prices charged by the State Electricity Boards (SEBs) from the various categories of consumers.
- Administered Pricing Mechanism (APM) to be dismantled in a short time-frame.
- Institutional reforms to be consolidated for deregulation etc.
- Optimum utilisation of existing assets.
- Production systems to be made efficient; transmission and distribution losses to be reduced.
- R&D, transfer of technologies and practices for environmentally sound energy systems, to be promoted.
- Energy efficiency improvement in accordance with national socio-economic and environmental priorities.
- Appropriate energy efficiency and emission standards to be promoted to reduce adverse impact on the environment.
- Labelling programmes for products to be established to provide decision makers and consumers with information on opportunities for energy efficiency.
- Adoption of energy-efficient technologies in major energy-intensive industries like iron, steel, chemicals, petroleum, pulp, paper and cement.

Following measures have been suggested for medium and long-term strategies:

- Demand management through greater conservation of energy; optimum fuel mix, structural changes in the economy; an appropriate modal mix in the transport sector, *i.e.* greater dependence on rail than on road for the improvement of goods and passengers, and a shift away from private modes to public modes for passenger transport.
- Shift to less energy-intensive modes of transport. Better urban planning would also reduce the demand for energy use in the transport sector.

- Shift away from depletable to inexhaustible resources, *viz.* solar, wind, biomass energy, etc.
- Exploitation of hydro-electric power, particularly for meeting peak demand.
- Emphasis on research, development, transfer and use of energy-efficient technologies and practices in the supply as well as end-use sectors.

Energy Sector Investment

Investments in Different Five Year Plans

Energy sector development since independence has been based on government support and public fund allocation. Power (electricity) sector was taken over by the government in view of its direct impact on industrial and agricultural growth. Subsequently, the coal and the oil sectors were also nationalised to meet the social and economic objectives of the country.

There has been priority channelisation of public investments to the energy sector programmes. Every Five Year Plan allocated up to 30 per cent of its funds for the energy sector; electricity sector receiving allocation support to the extent of 20 per cent of the outlay. Average annual investment in the energy sector was Rs. 772 million per year during the five year period 1951 56 which went up to Rs. 230 billion per year during the period 1992-97. The investment trends/plan expenditure during the five year periods from 1951-56 to 1992-97, in absolute figures and the share of energy sector investments in the overall annual public sector investments/outlays is given in the Table 1.7.

Energy Sector Investment in Post-1992 Era

After the liberalisation (in 1991) and with the opening up of economy to private sector, there has been reduction in the public investment to energy sub-sectors in the hope that the private sector will provide the necessary support. Private sector investment was, however, enough to meet the additional financial needs of the power and coal sub-sectors. As a consequence, during the five year period covering 1992-97, the investment in the power sector has declined to 2 per cent of GDP as against 2.4 per cent during the Sixth Plan (1980-85). Similarly, public investment in coal declined from 0.5 per cent of GDP in the Sixth Plan (1980-85) to 0.3 per cent of GDP in post 1992. On the other hand, investments in oil and gas exploration, oil refining and gas supply increased from 0.7 per cent of GDP in the Sixth Plan (1980-85) to 1.2 per cent of GDP in post 1992.

Table 1.7: Plan Expenditure and the Share of Energy Sector

Plan	*Period*	*Plan Expenditure (Rs. billion)*	*Share of Energy Sector (per cent)*				
			Power	*Oil/ Gas*	*Coal*	*NRSE*	*Total*
First	1951-56	19.6	—	—	—	—	19.7
Second	1956-61	46.7	9.7	0.8	1.9	—	12.4
Third	1961-66	85.8	14.6	2.6	1.3	—	18.5
Fourth	1969-74	157.8	18.6	1.9	0.7	—	21.2
Fifth	1974-79	394.3	18.7	3.6	2.9	—	25.2
Sixth	1980-85	1092.9	16.7	7.8	3.5	0.1	28.1
Seventh	1985-90	2202.2	17.4	7.3	3.2	0.3	28.2
Eighth	1992-97	4341.0	18.4	5.5	2.4	0.2	26.5
Ninth	1997-02	8592.0	15.6	6.5	2.6	0.4	25.1

Source: Ninth Five Year Plan 1997-2002; Planning Commission of India, 1999.

Energy Sector Reforms

The most recent series of policy changing in the Energy Sector were initiated in the year 1991 *vide* Industrial Policy Resolution announced in that year. The policy to promote private participation in the Energy Sector programmes is guided by the need for additional investment and competitiveness. Environmental management, protection and conservation of the natural resources has also emerged as a key priority in the reform process in the Energy Sector in functional terms.

Power Sector Reforms

The Power Sector Reforms included *functional reforms* in the form of investment promotion; *structural reforms* in the form of unbundling the existing vertically integrated monopoly of state electricity board into separate generation, transmission and distribution activities for rationalising their management; setting up of Central/National and State level Electricity Regulatory Commission etc. Various tax incentives, an assured rate of return to the investors have been granted. Liberalised tariff reforms for hydro electric projects have also been notified.

Coal Sector Reforms

These include permitting private sector in commercial coal mining by the required legislative amendments, offering coal blocks both for exploration and mining on competitive bidding basis; deregulation of coal prices; setting up of regulatory authority; re-structuring the coal sector by giving full autonomy to coal producing companies of Coal India Limited (CIL); and doing away with the concept of holding company; and accelerating the efforts for exploration of coal deposits etc.

Petroleum Sector Reforms

Functional Reforms in the form of investment promotion have concentrated on private participation in upstream and downstream activities; both refining and marketing operations were opened to private sector, a package has been formulated to attract private investment in the exploration of oil and gas which includes attractive fiscal concessions; a New Exploration Licensing Policy has also been announced to make exploration and production competitive, setting up of refineries in the private and joint sector have been permitted. The structure reforms are considered in the form of regulatory mechanism to a coordinated approach at the national level.

Future Energy Demand up to 2012

Energy Demand

The long-term demand for the end of Eleventh Five Year Plan (2011-2012) has been calculated. This demand corresponds to 6.5 per cent growth rate of GDP during the Ninth Plan (1997-2002), 7.7 per cent growth rate of GDP during the Tenth Plan (2002-2007) and 8 per cent growth rate of GDP during the Eleventh Five Year Plan (2007-2012). Table 1.8 gives the demand projection.

Table 1.8: Projected Demand for Final Consumption of Commercial Energy

	1994-95	*2001-02*	*2006-07*	*2011-12*
Electricity (Bkwh)	289.4	496.1 (8.1)	756.2 (8.8)	1150.2 (8.7)
Petroleum Pdts. (MMT)*	63.8	104.8 (7.3)	153.0 (7.9)	226.3 (8.1)
Coal (MMT)*	79.6	114.0 (5.11)	140.0 (4.4)	179.5 (5.1)
Natural Gas (MCM)*	12110	• 15730 (3.8)	18291 (3.1)	20853 (2.7)

Source: Ninth Five Year Plan (1997-2002) – Planning Commission of India, 1999

Notes: (1) Excluding demand for power generation.

(2) Figures in bracket are the CAGR over the period.

The installed generation capacity requirement in public utilities will be 31800 MW at the present level of capacity utilisation. This is 3.7 times the installed capacity of 85,019 MW in March 1997. This will mean an additional 233,000 MW in the utilities by the year 2012.

The installed capacity in 2001-02 works out to 130,763 MW requiring an addition of 45,744 MW, during the Ninth Plan period. The importance of this additional capacity is that 43.7 per cent is expected to be achieved through the private sector.

Primary Energy Production

The modal mix of power generation in future and the requirement of primary energy for the same will depend on the level of development of these modes in the coming years, the increase in the efficiency of power generation, transmission and distribution and end-use. The requirement of energy imports will increase but will be controlled to the extent by which the primary energy needs of the economy are met through indigenous production. Under the Business as Usual Scenario, the primary commercial energy production will be as indicated in Table 1.9.

The dependence on import may increase to 27.7 per cent by March 2002, and may go up steeply to over 47 per cent by March 2012 as against import dependence of about one-fourth in 1996-97.

The Planning Commission has developed three different scenarios for the import of primary energy to meet the projected energy demand as follows:

Table 1.9: Assumption for Primary Commercial Energy Production in BAU Scenario

	1994-95	*2001-02*	*2006-07*	*2011-12*
Hydro Capacity (MW)	20837	31456	34918	38380
Nuclear Capacity (MW)	2225	3105	4105	5105
Wind Generation (GWH)	182	1150	1900	2700
Solar Generation (GWH)	—	30	60	120
Crude Oil (MMT)	32.2	37.0	40.0	45.0
Natural GAS (MCM)	17339	29165	33915	38665
Coal (MMT)	253.7	370.6	498.9	671.8
Lignite (MMT)	19.1	46.1	61.0	75.0

* Natural gas production is net of flaring which is assessed at 5 per cent of gross production

Source: Ninth Five Year Plan (1997-2002) — Planning Commission of India. 1999

Scenario I: Business as usual, *i.e.*, no change in pattern of use.

Scenario II: Accelerated hydro power development in the next 15 years, thereby increasing the share of hydro electricity capacity in total installed generation capacity to 25 per cent, 30 per cent and 35 per cent in the year 2002, 2007 and 2012 respectively, as compared to nearly 20 per cent in 1997; and liquid fuel based power generation capacity addition at the rate of 8,805 MW for Tenth Five Year Plan (2002 to 2007) and Eleventh Five Year Plan (2007 to 2012).

Scenario III: Savings of 10 per cent in electricity and oil consumption to be achieved in the period Tenth Five Year Plan (2002 to 2007) and Eleventh Five Year Plan (2007 to 2012) as compared to a saving potential of about 7 per cent in the Ninth Five Year Plan period, 1997-2002.

Energy Imports

The estimated net imports of coal and liquid fuels given in Table 1.10.

Issues for Future Sustainable Energy Supply

Population Dynamics

The major issue or the major concern that will dictate future course of sustainable energy supply is the population.

Table 1.10: Net Import Requirement of Commercial Energy

	2001-02	*2006-07*	*2011-12*
Coal (MMT)			
Scenario I	40.5	161.7	388.5
Scenario II	29.8	45.2	101.9
Scenario III	(-) 3.5	(-)17.1	11.4
Oil and Oil Products (MTOE)			
Scenario I	75.2	137.2	212.4
Scenario II	75.2	145.6	228.8
Scenario III	68.3	127.5	205.0

Note: Negative net imports implies surplus availability from domestic sources. However, some coal imports may be required for specific uses.

Source: Ninth Five Year Plan (1997-2002) — Planning Commission of India, 1999.

India touched a population of 1 billion on May 11, 2000, *i.e.* 16 per cent of the world's population on 2.4 percentage of the globe's land area. If the current trends continue then India may overtake China in 2045, to become the most populous country in the world. The growth in population during the decade March 1991 to March 2001 will be from 846.3 million to 1012.4 million. The projections made by the National Population Commission in 1999 indicate that India's population at the end of 2020 would be around 1350 million if the present trends continue. However, if the National Population Policy 2000 Action Plan fail to implement, the population may be around 1450 million in the year 2020. Whatever be the case, there is substantial increase in the population that needs to be addressed.

It is estimated that presently (1999) about 32 per cent of the population is in the urban areas and 69 per cent in the rural areas. By the year 2020, almost 50 per cent of the population will be in the urban areas. This simply in terms of number means that by the year 2020, the urban population would have jumped from present 310 million to over 1675 million and the rural population would have reduced from 675 million to 600 million.

Urban Energy Needs

The major issue of concern, therefore, will be to ensure higher levels of electricity availability to the urban population. Since they will be able to afford to buy electricity, it will only be an issue of electricity or energy availability to them. The preferred form will be quality, reliable power supply but that may not be available. Most of this power supply will come only from major base load power plants.

Rural Energy Needs

Out of about 600,000 villages in India 80,000 villages are not electrified. Of these, 18,000 villages may never be electrified from grid. It is estimated that about 500 million people in India are without access to commercial forms of energy, barring kerosene being used for small lighting. This reflects the situation of energy poverty. Their dependence, mainly on the biomass fuel used in the traditional form, has to be thoroughly studies and subsequently efforts are to be made to improve efficiency of use and conversion.

Inappropriate use of solid biomass fuels as the only medium of cooking energy indoors has resulted in most localised and health damaging pollution. It has been observed that with the rapid urbanisation, the cooing have been transferred to the urban centres also; and urban slums are becoming a major source of pollution from such activities.

Reaching out electricity to 675 million people will be the major challenge for the country. It is expected that that the income levels of these rural power will improve as poverty alleviation measures will have a more positive impact due to rapid impact of inclusive globalisation (a term suggested by the Secretary General of UN in his Millennium Report) whereby the process of globalisation will be centred round the rural people.

In the coming decades, the cost of finance will be increasing as it will come from the private sources who will have more competitive use or better returns from the other sectors. Thus expecting rural energy supply, based purely on the market applications, is ruled out. The government is making provisions in the proposed Electricity Bill, 2000 to ensure electricity supply by public intervention by providing the sunk cost of generation equipment.

The renewable policy, now on the anvil, also proposes to substantial improvement in rural energy supply by decentralised power generation from biomass, wind and solar sources. At the same time efficiency improvement in equipment will be of extreme value for the villages.

Efficiency on Supply and Demand Side

Policy support for efficiency improvement and conservation has been persuaded for the last three decades. It was only during late 80s when specific steps were taken for ensuring efficiency improvements on the supply side. The impact was that the outdated power plants undertook steps for introducing retrofits to recondition and improve their performance. A substantial improvement in the power load factor resulted additional electricity in the supply grid to a value equivalent to 5,000 MW from the same installed capacity. That was also considered a major element in meeting energy needs of the country during the period 1992-97 when GDP growth rate of 6.8 per cent could be attained.

End Use Efficiency Improvement of Capital Goods

The economic reforms also lead to introduction of new and modern capital goods in the manufacturing sector, thereby bringing in efficiency improvement. ISO 9000 and ISO 14000 have now become a norm rather than an exception. These steps are expected to lead to improvements in efficiency of operation and subsequently improvement in efficiency of energy consumption.

Quality and Reliability Power Supply

Erratic power supply in the country has often caused disruption in operation and forced commercial and domestic users to opt for additional alternative power generation equipment. It was estimated that during the year 1998-99 the total quantity of 5 million tonnes of crude oil equivalent was used as a swing fuel to overcome such erratic stoppages. Improvement in performance of power plants and availability of electricity in abundance is likely to reduce the consumption of swing fuel. There may be a necessity for the government to intervene and enforce restriction on use of alternative power generation devices.

Rational Transport and Traffic Policies

The Government of India has finalised an Integrated Transport Policy that has taken a rational overview of energy consumption per tonne of freight movement and thereby decide the optimal mode of transport between rail and road. At the same time emphasis on infrastructure mainly on road, rail and port is expected to smoothen the traffic flow, leading to lower energy consumption. Metro cities like Bombay, Chennai, Bangalore and Hyderabad have already undertaken massive traffic movement regulation efforts and built a number of flyovers and new arterial roads. The city of Delhi has launched a massive programme of 32 flyovers to be completed by 2002; and the underground rail system to smoothen the traffic and reduce the number of new vehicles. That will lead to energy efficient and more sustainable transportation system.

Regional Cooperation

India has planned a 'national grid' with connectivity with Bangladesh, Nepal and Bhutan. Future possibility of massive hydro power resource utilisation in Bhutan and Nepal can help the economies of collaborating countries when the electric power can be appropriately evacuated through the regional grid. At the same time, the possibility of trade in natural gas between Bangladesh and India also holds promise, but will depend on the local requirement of natural gas in Bangladesh.

Financing of the Projects

Major energy sector projects to be developed by private and international funding call for risk coverage to attract foreign investments. India has shown its interest in the joint venture applications to assure the investor of their returns. Regional joint venture can certainly help in more assured sustainable energy supply in the country.

Conclusion

Major initiatives undertaken by the Government of India in terms of regional cooperation; integrated transport policy; development of environment friendly renewable sources including the hydro; and energy efficiency and conservation measures on supply and demand side can lead to sustainable energy supply in the country. Necessary legislative and administrative measures have now been developed to assure foreign investors of all available support systems to provide risk cover to their investments.

REFERENCES

Chaturvedi, Pradeep, (1999). ***Energy Management***; Concept.

Economic Survey of India (1999). Government of India.

Ministry of Coal, Government of India, Annual Report 1998-99.

Ministry of Non-Conventional Energy Sources, Government of India, Annual Report 1998-99.

Ministry of Petroleum and Natural Gas, Government of India, Annual Report 1998-99.

Ministry of Power, Government of India, Annual Report 1997-98, Annual Report 1997-98, Annual Report 1998-99.

National Paper (1998). *India's Energy Scenario 1998*, Presented at the 17th Congress of WEC at Houston, September 13-18.

Ninth Five Year Plan (1997-2002). Government of India, 1999.

2

Energy Vision for Sustainable Growth

Introduction

The prime challenge before the country is to provide the minimum energy services to allow the poor people to achieve a decent standard of living. By 2020, which is also the target goal for the Vision Statement, the provision of modern energy services can be ensured either as a connection to a reliable, sustainable electricity grid supply, or as stand alone electricity supply system.

It is necessary that the Energy Vision 2020 focuses on the following three issues:

(i) Overcome energy poverty wherever it occurs.
(ii) Enhance the quality and reliability of delivered energy at affordable price; and
(iii) Minimise the indicative environmental and health impacts of energy development.

Commercial Energy Sector Growth

Based on end-use analysis and past income elasticities, the Planning Commission estimated that the requirement for primary energy is likely to increase from 374 mtoe at the end of the Eighth Five Year Plan (1996-97) to around 475-500 mtoe by the end of the Ninth Five Year Plan (2001-2002), i.e. an annual growth rate of 4.9 per cent. Of the share of commercial energy is likely to increase from 66 per cent at the end of Eighth Five Year Plan to 75 per cent at the end of Ninth

---

Paper presented at All India Seminar on Power for Sustainable Growth, Shillong organised by the Institution of Engineers (India) Shillong Local Centre during August 2001 at Shillong.

Five Year Plan, resulting in corresponding decline in consumption of non-commercial energy. Thus, the growth of commercial energy will be 6.8 per cent.

Energy import dependence is also likely to increase 25 per cent at the end of Eighth Five Year Plan to 28 per cent at the end of Ninth Five Year Plan. In absolute terms it means that our energy imports will increase from about 95 mtoe to over 140 mtoe, which is a quantum jump of over 50 per cent.

Total additional power generation capacity installed during the Eighth Five Year Plan period (1992-97) was only about 16,422 MW as against the planned 30,538 MW. Still GDP growth rate of 6.8 per cent could be attained. A number of thermal power plants exercised retrofits and better maintenance practices, and their efficiency of operation and plant load factor improved significantly. Higher output from existing power plants has partly compensated for lower additional capacity. In addition, improvement in end-use efficiency in the industrial sector due to induction of imported latest capital goods also reduced demand for final energy.

Erratic power supply and low voltage often forced end users/ consumers to opt for additional sources of power generation, viz., diesel/ kerosene/petrol generating sets. It is estimated that whereas in 1993-94, one mtoe of diesel was used as swing fuel the consumption in 1997-98 has gone up to about 5 mtoe.

The status of different sectors of primary energy is also disturbing. Three observations are relevant:

(i) Slippage in power capacity augmentation.
(ii) Lower off-take of coal.
(iii) Stagnation in domestic production of petroleum products.

The Ninth Plan mid-term appraisal by the Planning Commission has projected lower achievements in capacity additions in the power sector. During 1997-2000, capacity to be added was projected at 24,309 MW against the total of 40,245 MW for the whole plan period. During the first 3 years of the plan period (1997-2002) only 12,035 MW have been realised due to inordinate delay in implementation of new ventures by private sector companies.

The Ninth Plan (1997-2002) projects that 17,588 MW of capacity would be added by the private sector. The current estimate is that private projects may add only 7,702 MW. States are expected to realise capacities of 9,893 MW.

In case of coal off take during the Ninth Plan, the projected figures will not be achieved as the growth rate has slowed down and the power capacity additions have not been realised. Government's permission to import coal at zero per cent duty by the cement sector has increased imports and affected local coal consumption. The Planning Commission has slashed the growth target in coal off-take to 2.86 per cent against 5.34 per cent as projected earlier. Accordingly, the coal production target has been lowered to 328.86 mt against the initial estimates of 371.6 mt in the year 2001-2002.

The mid-term appraisal also observes that in the petroleum sector the exploratory work has not identified any additional oil and gas reserves. There is an urgent need for larger exploration to boost domestic oil production

Failure to set up meet additional power generation capacity resulted in lower consumption of coal at 310 mt; and higher consumption of oil and oil products at 102 mtoe in 1999-2000.

On the positive side the performance of nuclear power plants has been heartening. After remaining dormant for a number of years, the nuclear power plants operated at PLF of over 80 per cent in 1999-2000. NPCIL power plants generated 13,394 MUs in 1999-2000 against 7,983 MUs in 1995-96. Simultaneously with the improvement in generating performance, a very high level of safety has been maintained. Judging by recent performance nuclear power is geared for stable power supply.

Foreign and Private Investment in Energy Sector

India has so far attracted private and foreign participation only for power generation sector sufficient to set-up about 5,000 MW capacity (since 1992); and concrete proposals for another about 5000 MW are on the anvil.

Oil and gas exploration efforts have been geared up through award of 25 exploration blocks to private/joint/foreign initiative under the New Exploration Licensing Policy (NELP). The Production Sharing Contracts have been finalised for 22 of 25 blocks. Twelve new oil fields have been discovered. The private investment is still to flow in production capacities.

Coal mining efforts have not picked up in the private sector as the present provisions are only for mining of captive use. Bill is pending in the Parliament for its approval for unrestricted private entry for exploration and production of coal. After amendment to the Coal Mines Nationalisation Act 1973, the Coal Bearing Areas Act 1957 will also have to be amended so as to provide a level playing field to the private sector in the matter of issuing prospecting/mining licenses. At present only six blocks have been allocated to the private sector and coal mining at a level of 2 tonnes per annum is reported.

Low level of foreign direct investment and private investment indicate that India has still to establish legal, financial and market reforms which will attract the necessary domestic and foreign capital for new energy projects. Higher risks and transaction costs are generally leading to expensive investments.

A fundamental shift is that the primary energy markets, mainly in the oil and gas sectors, are being taken out of the administrative regime (to a certain extent). It is expected that by the year 2002 the primary energy would be mainly traded through market.

Besides the commercial financing, the other channel of support to developing country energy market is through overseas Development Aid. Financial cooperation between developed and developing countries has not shown any improvement in the reforms period (since 1992). According to recent World Bank data, OECD countries have failed to fulfill their international commitment of official Development Aid (providing no more than 0.2 per cent GDP in 1998 compared to 0.7 per cent promised by them at the World Social Summit at Copenhagen in June, 1995). This had perceptible effect on India. Also the Pokharan Nuclear Test of May 11, 1998 became an excuse for a number of countries to stop aid to India. Therefore, the ODA to India in the energy sector has declined.

Characteristics of Energy Sector Growth since Introduction of Reforms in 1992

Following characteristics of growth have been observed in different sectors of energy (in September 2000) :

(a) *Additional power generation capacity target not achieved* : Power generation has not developed as envisaged, mainly due to lack of enabling environment for foreign and private investment.

(b) *Low coal consumption and production*: Coal consumption and production has been lower than projected due to lower additions of power generation capacities.

(c) *High oil consumption and import*: Oil consumption has been high, as a large number of captive power plants were installed and larger transport activities took place and service sector developed. Import of crude oil increased as domestic production stagnated.

(d) *Natural gas use on the increase*: Natural gas use has been on the increase. Combined cycle gas turbine, utilising natural gas, for base load, has been projected as a potential front-runner in the country for future additions of power capacities. Option of Naptha and LNG based power plants have also been considered. CNG has been introduced in the transport sector. Lack of infrastructural facilities for local transportation/ transformation, mainly the port facilities, is delaying decision.

(e) *Low growth of renewables*: Contribution of new and renewable sources of energy has been lower than projected, except for biomass (in traditional form) and wind power. Lack of policy on non-renewables and hidden subsidies for conventional sources have been mainly responsible for the same.

(f) *Nuclear power growth significant*: Nuclear power has registered significant growth, mainly due to rehabilitation of the Rajasthan Atomic Power Plant. Systematic work on safety, waste management and environment management have raised level of public confidence. Environment management control have earned the ISO 14000 certificate to Narora Atomic Power Plant.

(g) *Primary energy import on the increase*: Dependence on imported primary energy has increased, even though the refining capacity has gone up to 102 mt per year.

(h) *Pollution abatement programmes initiated*: UN Framework Convention on Climate Change, in 1992; and Kyoto Protocol, in 1997, have resulted in wider awareness of environmental issues relating to energy use. Local and regional pollution as well as greenhouse gas emissions have received wide political attention. Adverse impact of greenhouse gas emissions, on air pollution and health, has been under scrutiny. Abatement plans and programmes are initiated and results are seen in the energy sector.

Energy Demand 2020

A few model studies have been presented by experts in the recent past for projecting energy demands for the period upto 2020. The Study conducted and projections made by E. A. S. Sarma, J. N. Maggo and A.S. Sachdeva has been found nearest to the ground realities, and adopted by us in our practice.

The Study indicates overall energy demand perspectives for three different scenarios as follows:

(i) Referenceable Scenario (BAU) — in which the business as usual would be allowed to continue.
(ii) Efficiency Oriented Scenario (EFF) — considering the scope for reduction in the energy intensity based on induced efficiency through technology and management adaptations; and
(iii) Environmentally Constrained Scenario (ENV) — considering the implications for emphasis on minimisation of adverse environmental impacts in augmenting the energy supply system.

Energy Demand and Sector-wise Break-up – Business as Usual (BAU) Scenario

In a presentation in 1996, the authors made projections under the "business as usual"(BAU) as that was considered the referenceable scenario. Total demand by 2020 worked out to 39,107 peta joules (934.0 mtoe) as compared to 9,727 peta joules (232.3 mtoe) in 1989 (corresponding figure for 1997-98 is 390 mtoe).

Expert consultations agreed on energy demand projections (under BAU) for India as given in Table 2.1.

In order to meet the level of final energy demand projected in the BAU Scenario, the installed capacity in power generation will need to increase to nearly 3,85,000 MW in the year 2020. The coal and oil requirement will correspondingly be 1,400 mt and 335 mt respectively.

Table 2.1: Future Demand for Commercial Energy in India (BAU Scenario)

Source	Unit	*1991-92*	*1996-97*	*2009-10*	*2020-21**
1. Electricity	TWH	231	336	725	1350 (1745)**
2. Coal	mt	229	311	690	1400
3. Petroleum Products	mt	57	81.2	165	335
4. Natural Gas	bcum	18.6	30.2	65	130

Source: Planning Commission, Government of India — September 1995 and September, 1996 and India's National Paper on Energy Scene 1998 presented at the 17th Congress of the World Energy Council, at Houston in September 1998.

* *Projections to 2020-21 made on the basis of 'Business as Usual' — J.N.Maggo — R&D Needs for Energy Sector — September, 1996*

** Electricity generation capacity of 1745 TWH will be required to take care of T&D losses and Plant Load Factor. Corresponding installed generation capacity will be 385,000MW at present day PLF and T&D loss level (1996).

Demand for non-conventional energy sources, mainly biomass fuels will remain around the same figure of about 130 mtoe (as ii 1998-99) due to interfuel substitution and population migration from rural to urban areas.

Energy Demand and Sectorwise Break-up – Environment and Demand Side Efficiency Driven Scenario

Globalization and competitiveness of Indian economy is creating a scenario wherein efficient and cleaner technologies will be introduced. Introduction and adaptation of such technologies will be slow for reasons of outdated equipment and plants and lack of financial capital to change them fast. Therefore, it is expected that the combined impact of these measures can lead to at least 10 per cent lowering of energy demand in 2020, compared to BAU Scenario (except gas). Thus applying this assumption to BAU scenario, the commercial energy demand will work out to about 36,000 peta joules (850 mtoe).

Thus the sectorwise energy demand projection under environmentally sound and, efficiency driven use scenario will be as given in Table 2.2.

Correspondingly the final energy demand in environmentally sound and demand side efficiency supported scenario will need an installed capacity in power generation of 3,50,000 MW in the year 2020. The coal and the oil requirement will correspondingly be 1,260 mt and 300 mt respectively. The natural gas demand will be 130 bcum.

Table 2.2: Future Demand for Commercial Energy in India (Env & Demand Side Efficiency, EFF Scenario)

Source	Unit	1991-92	1996-97	2009-10	2020-21
1. Electricity	TWH (1570)	231	336	725	1215
2. Coal	mt	229	311	690	1260
3. Petroleum Products	mt	57	81.2	165	300
4. Natural Gas	bcum	18.6	30.2	65	130

Demand for non-conventional energy sources is expected to remain at 130 mtoe.

Energy Demand and Sectorwise Break-up – Environment, Demand Side and Supply Side Efficiency Driven Scenario

If the supply side power sector interventions are taken into consideration, then the advantage of PLF improvement and T&D loss reduction will reduce the power generation capacity requirement by almost 10 per cent. Correspondingly the installed capacity required in 2020 will be 3,20,000 MW for a delivered energy supply of 1215 TWH/year from 1,400 TWH/Year generated power. Correspondingly the coal requirement will come down to about 1150 mt. However, a number of future power plants are to be based on liquid or gas fuels. Some coal experts project that the demand of coal can reduce to just 800-810 mt if power generation from hydel, nuclear and renewable sources of energy can be organised appropriately. Thus, taking into consideration the above facts the primary commercial energy demand in 2020 can be seen in Table 2.3.

The demand for non-conventional energy sources is likely to be 140-150 mtoe in view of power generation from biomass.

Suggestions for the Power Sector

Future demand for commercial energy envisages additional installed power generation capacity to the tune of 210,000 MW to the existing capacity of over 108,691 MW as on March 31,1999. The energy demand projection indicates that the total power requirement in 2020,

Table 2.3: Future Demand for Commercial Energy in India (Env & Demand Side and Supply Side Efficiency Scenario)

Source	*Unit*	*1991-92*	*1996-97*	*2009-10*	*2020-21*
1. Electricity	TWH	231	336	725	1215
	(1400)				
2. Coal	mt	229	311	690 (810)*	1150
3. Petroleum Products	mt	57	81.2	165	300
4. Natural Gas	bcum	18.6	30.2	65	130

* Coal requirement will be 1150 mt in 2020, if the power generation from thermal power will be 75 per cent of the total. However, since contribution of nuclear power at 20,000 MW; renewable energy power at 20,000 MW and power import from Nepal and Bhutan at 20,000 MW are also considered, the requirement of thermal power will reduce to 190,000 MW (about 110,000 MW additional power over the present capacity of about 80,000 MW thermal power in utilities and non-utilities combined) or about 60 per cent of the required installed capacity of 320,000 MW, and the overall coal requirement will be only about 810 mt. Some part of additional thermal power will come from gas and liquid fuels.

under business as usual scenario (BAU), will be about 280,000 MW and the corresponding installed capacity requirement will be 385,000 MW considering the plant load factor and transmission and distribution losses as they existed in 1994. Internalising the environmental factors and introducing demand side efficiency measures are expected to bring down the installed generation capacity requirement by 10 per cent to 350,000 MW. However, the new plants have shown a consistently high PLF of over 90 per cent; and many of the existing plants have been retrofitted and made operational at over 80 per cent PLF. At the same time the transmission and distribution losses have gradually come down and with the installation of more stable grid and installation of sophisticated instruments and equipment on the grid the transmission and distribution losses will also be reduced. The combined effect of both these can result in improving supply side efficiency by 10 per cent thereby reducing the installed generating capacity requirement by 10 per cent. Energy needs can be met in 2020 by installing only 320,000 MW power generation capacity; or an additional of only 210,000 MW of power generation capacity in the country by the year 2020. This will be possible only under demand side and supply side efficiency driven scenarios.

The suggested mix of power supply in 2020 is represented in Table 2.4.

Table 2.4: Power Sector Supply Pattern to meet Demand Projection for 2020 *(in MW)*

	Hydro	*Thermal*	*Nuclear*	*Wind and other Renewables*	*Imported Power*	*Total*
Installed generation capacity as on 31.3.1999						
-Utilities	22,438	68,586	2225	942		94,191
-Non-Utilities		14,500				14,500
Total	22,438	83,086	2225	942		108,691
Projected Installed Generation capacity on 31.3.2021	65,000*	195,000	20,000	20,000	20,000	320,000
Additional Generation Capacity to be installed 1999-2020	42,562	111,914	17,775	19,058	20,000	211,309
Total						320,000

* The contribution of hydro will be 25 per cent of hydro & thermal combined. figures have been extrapolated.

Power Sector Initiatives Agreed at the Chief Ministers' Conference

The Prime Minister addressed the Chief Ministers' Conference on March 3, 2001 to work out if the power sector reforms can be put in place. Referring to 1996-98 and 2000 high-powered Committee and Chief Minister's meetings, the Prime Minister emphasized on Chief Ministers to look into the earlier recommendations and ensure power sector reforms to promote commercial viability of State Electricity Boards in two years. The 'consensus' to follow PM's suggestion was considered a 'landmark decision' and could be achieved only because of a shift in approach of the central government to be 'accommodative' and let the process of reforms evolve rather than enforce. The central government agreed to provide all necessary funds to facilitate such states who opt for reforms. In return, the States have agreed to make users pay for power and put an end to the financial drainage on State Electricity Boards. Every state government was given freedom to choose its own mode of reforms which does not necessarily have to be unbundling of State Electricity Board.

Some of the earlier decisions were reiterated and have been now agreed. These include the following:

(i) Every State to rationalize tariff within six months through independent State Electicity regulatory Commission. (States will have the freedom for a joint SERC for a number of states).

(ii) Transparent policies on subsidies to be provided to the customers only through their budgets and not through tariff.

(iii) Benchmarking of tariff at a minimum of 50 per cent of the cost of supply and agricultural tariff not be less than 50 paise per unit.

(iv) Energy audit to be set up within six months at all 11 kV feeder points throughout the country.

(v) 100 per cent metering of consumers in the country in two years.

(vi) All power thefts to end in two years.

(vii) Committee of Experts, from Centre, State and Central Power Utilities to work out an agreement within three weeks for one time settlement of Rs.27,000 crores owed by State Electricity Boards to central power utilities. Ongoing dues to be cleared as they arise, if needed by deducting directly the money from State's share from central funds.

(viii) Each State to work out a two year schedule to make State Electricity Boards commercially viable.

In addition, the budget for the year 2001-02 outlines that states committing themselves to a time-bound reform plan will get the first claim on money from the accelerated power development programme and fiscal reform incentive fund. At the same time, existing offer of five year's tax holiday for firms investing in generation, or in a new network of transmission/distribution to be extended to any of the first 15 years if investment is in place before April, 2006.

Salient Features of Draft Electricity Bill 2000

The budget proposals for 2001-02 indicate that the Finance Minister has fully identified the problems, which are basic to reforms. But, he feels that he could get these reforms cheap. He is expecting that by providing lower interest rates and incentives of tax holidays the private investor will take the responsibility of supporting reforms. All his second-generation reforms require commitment of public funds to compensate the losers, to provide cushions for contingencies and insurance against uncertainties. The Finance Minister is operating in a poor economy where public funds are scarce and social services needing public allocations are gigantic. Also under the democratic set-up the feasibility of any policy is dependent on its acceptance by the people. They must perceive that the policy serves their interests today, and if not today, in not too distant a future.

To facilitate reforms in the power sector, certain amendments were introduced in the Electricity Act. However, those did not provide a total package of legislative measures to facilitate reforms. The Ministry of Power, therefore gave the responsibility to National Council of Applied Economic Research to draft the Electricity Bill 2000.

Subsequently, the Draft had to be revised several times till the Eighth Draft was submitted in September 2000 to the Ministry of Power. This Draft has been studied by the government and suitably modified to make it acceptable to the Parliament.

The Ministry of Power arranged a meeting of experts and officials on March 9, 2001 to discuss the Draft Bill. Salient points of the Draft Bill have been indicated as follows (detailed under different categories):

Generation

- Generation will be free from licensing. New Generation projects would not require techno-economic clearance from CEA. However, they would need to conform to the minimum technical standards laid down by the CEA.
- Setting up of captive generation plant would be free.
- Hydro Projects would need clearance of the State Government. The State Government would obtain prior consent of the Central Government from the view of impact on riparian rights of any other States.

- Promotion of generation from non-conventional energy sources as well as co-generation will be encouraged by the Regulatory Commissions.

Transmission

- Load Despatch to be done by an independent entity not being a private entity. Its directions to generating stations, transmission companies as well as distribution companies would be binding.
- There would be a transmission utility for managing and developing the transmission system to ensure that transmission is not a bottleneck in carrying power from generating stations to load centers in an efficient and optimal manner.
- Private transmission licensees for specified transmission lines would be licensed by the concerned Regulatory Commission after due consideration of the views expressed by the transmission utility.
- No trading in power by the Load Despatch Centre/ Transmission Utility/Transmission Licensee (as is the international practice).

Distribution

- Distribution would need licensing.
- Retail tariffs would be determined by the Regulatory Commissions, until competition is established, when only wheeling charges would be regulated.
- Metering made mandatory within two years.

Rural Areas

- State Governments and State Regulatory Commissions to endeavour completion of rural electrification.
- Formulation of a national policy in consultation with State Governments and SERCs for bulk purchase of power and management of local distribution through the Users Association, Cooperative, Franchisee or Panchayat Institution

etc. for rural areas.

- Non-conventional energy sources and stand alone systems would be freely permitted.

Trading, Market and Open Access

- Supply and trading of power would be a distinct activity and would require licence.
- Regulatory Commission to promote trade and development of a market including a spot market subject to regulations
- Regulatory Commission to permit open access to the distribution lines in phases to promote competition and efficiency.
- Consumer tariff to progressively reduce cross subsidies and move towards actual cost of supply.
- State Government to provide subsidy through the budget for specified target groups if it requires the tariff to be lower than that determined by the Regulatory Commission.

Restructuring

- Conversion of existing SEBs into one or more company within a year through a transfer scheme which would provide statutory transfer of assets, liabilities and staff. However, if a State wishes to continue with the SEB beyond this period, it can do so through a notification.
- The service conditions of the staff of SEBs would not be inferior consequent to their transfer to the new company.
- Creation of a State Regulatory Commission would not be mandatory.
- Two or more states/UTs or a combination of both may have a Joint Electricity Regulatory Commission, if they wish to do so.
- Grants for Regulatory Commission would be voted. They may allow to retain fees prescribed with the approval of the Government to meet their expenses.

CEA

- CEA would continue as the main technical advisor of the Government of India with responsibility for power planning. The CEA would also be the technical advisor to the Central as well as State Electricity Regulatory Commission.

Policy Actions

It is necessary to look at the basic requirements for overall socio-economic development and alleviation of poverty in India to arrive at the agenda for energy development. Sincc India's emerging needs are greater than the indigenous resources, it is important to access regional energy resource through trade.

Energy sufficiency in 2020 can be attained through:

(i) indigenous energy resources of coal, oil, gas, hydro power, nuclear power, and new renewable sources of energy including biomass;
(ii) oil and coal import from major international markets;
(iii) hydro power import from Bhutan, Nepal; and
(iv) gas import from Bangladesh.

Following policy actions will be desirable for attaining the goal of energy sufficiency in 2020:

Affordable Energy for the Poor

To make energy affordable to the poor, the government will have to shoulder the responsibility for absorbing part or all the sunk cost of energy infrastructures; support dencentralized energy systems and build the capacity of local energy enterprises.

Speedy Implementation of Market Reforms

The introduction of Electricity Bill 2000 in Parliament is aimed at supporting market reforms. Almost a decade has passed but the market reforms are still at the discussion stage. The Government needs to allow a free hand to state governments to choose their models. Economic considerations will dictate the model.

Government's Role in Energy Sector

Government will have to continue to play a dominant role in the energy sector by providing public funding for atleast over the next 20 years to facilitate and promote the process of privatisation.

All Energy Options are to be Effectively Developed

Recent advances in the nuclear power sector and development in renewable energy sector indicate that all the sources can significantly contribute to the additional energy requirements over the next 20 years. They should be appropriately utilized.

Ensure an Enabling Climate for Investment in the Energy Projects

The Government will have to ensure that all regulatory frameworks are in place at an early date and reasonable profits are allowed to future investors.

Energy Prices to Cover Cost and Ensure Payment

End use prices are a key parameter driving energy consumption in the developed countries. It is not so in case of developing country like India. Unless these costs reflect long term marginal costs including wherever possible the cost of well identified externalities like energy security or environment protection. Metering is essential to ensure payments for the electricity consumed by users.

Encourage Energy Efficiency

Energy Efficiency policies that use direct or indirect price mechanisms are the most effective in lowering energy consumption trends. Legal standards and an adequate payment system for energy are essential to achieve energy efficiency goals.

Provide Larger Fund Support for Research, Development and Deployment

As R&D in the energy sector cannot be made critical only through market mechanism, there is no incentive for the private sector to invest in equipment and product research. Such research will have to be supported largely through public funds either to the public institutions or to private sector organizations.

Promote Regional Trade in Energy

Long-term collaborative projects should be encouraged to tap hydro resources in Bhutan and Nepal and gas resources in Bangladesh. The combined effect of regional trade will be valuable for the development of North East and East India.

Link Financing Partnerships to Environmental Goal

Given the enormous need for new capital stock in developing countries, international mechanisms with the potential to stimulate capital flows linked to clean and safe energy projects in developing countries are valuable supplementary approaches which should be a high priority of governments.

Use Skilled Manpower from Existing Projects

Due to completion of ongoing projects and lack of new opportunities the skilled manpower, mainly in hydro projects, is languishing under different agencies. Government should develop a policy structure in disinvestment in the energy sector in such a fashion that joint sector participation is promoted whereby the experience of skilled manpower can be put to effective use.

Ensure Energy Security

The option to enhance energy security include; avoiding excessive dependence on imports, diversifying supply, encouraging technology transfers and increasing national and regional and strategic reserves of crude oil and oil products through increased investment and advanced exploration technologies.

Conclusion

Power generation capacity can be ideally added in environment friendly fashion if the hydro resources contribute to 1,25,000 MW or the 35 per cent required installed capacity of 3,20,000 MW in 2020. In the present context of environmentalists' uprisings it is difficult to set-up hydro projects to that capacity.

However, a 60,000 MW supplement from nuclear power renewable based power and power import will call for only 65,000 MW from hydropower to attain the goal of environment friendly power production. Thus emissions from the energy sector will be substantially curtailed and correspond to the stage of 35 per cent contribution of hydropower resources.

Nuclear power will play a pivotal role for the base load supplies in the year 2020.

3

Quality Power Distribution Through Private Companies

The Power Reform Process in the country introduced in the early 90s has indicated that the process has really not made its impact in the Indian context. Basing its approach on other countries experiences the Government of India focused attention on exploiting foreign capital and private capital into the power sector. The yardsticks used by the developed countries were followed as indicated to the road map for power sufficiency in the country. It was subsequently realised that various elements that is helping the market approach in the developed countries can be considered, leave aside introduced, if only the overall market situation is fully developed. The World Energy Council came out with a study on financing of Energy Sector in 1997, which clearly indicated that barring about 25 per cent of the total investments, the foreign companies will try to tap only the local or the domestic capital markets for the funds required in the power sector in India. The bitter experience of the initial phase of power sector reform made government to have introspection on the earlier approach and the alternate development strategy. As acceleration in economic growth depends heavily on the energy sector, the power sector cannot be ignored for long time. The Finance Health of the State Electricity Boards has become a matter of grave concern, considering that their losses have reached an alarming level of Rs. 26,000 crores, which is equivalent to above 1.5 per cent of GDP. World Bank estimates, this loss would increase to Rs. 40,000 crores in the next five years, unless this trend is halted with corrective steps. Out of total energy generated, approximately on 55 per cent is build and only 41 per cent is realised. The gap between average revenue realisation and average cost of supply

---

Keynote Address at the International Seminar on Power Sector Distribution Reforms in Delhi organised by Chem Tech Foundation in November. 2002 at New Delhi.

has been constantly increasing. During the year 2000-01, the average cost of supply from SEBs was 304 paise per unit and average revenue realisation was 212 paise per unit.

Realising that in addition to lack of capacity to attract foreign private capital, huge T&D losses and lack of accounting and accountability in distribution have adversely affected the financial health of SEBs.

The government has developed a six-level intervention strategy aimed at improving the sub-transmission and distribution network, that cover the entire country in the next three to five years. The accelerated power development programme, launched in February 2001 is an exercise aimed at renovation and modernisation of old power plants and upgradation and strengthening of sub-transmission and distribution networks including energy accounting and metering in the distribution circles in a phased manner.

On the eve of launching of the Tenth Five Year Plan, it was realised that improvement in employment creation demands higher growth rates that can only be achieved through better infrastructure facilities. These better infrastructure facilities, however, are dependent on availability of electric power. IT and communication are considered two pillars for sound economic development. But both of them are totally and absolutely dependent on the quality of power, made available for industrial development. The distribution network plays a very important role. This network will include introduction of the latest of the IT based enablers to ultimately enable IT sector and the industry to go.

Twenty years ago, utility customers were not so demanding. Utilities could take the customers out of service to work on the line and no one thought the worse. Customers appreciated a notice that the power would be out, but they certainly did not expect it. Of course, with high-tech businesses uses, fancy digital controllers, and increasing computer loads, power outrages now mean lost incomes. Residential customers, and remote operating professionals, with their home computers and their desire for convenience and comfort are no longer accommodating.

While customers would prefer to have their electricity delivered free of imperfections, they tolerate some outrages better than others. The worldwide qualify and prices are rated high in importance but lower in satisfaction. Statistics have further broken down into categories including troubleshooting, billing, and customer responsiveness and claim settlements.

Discussing about the role of better distribution system, it is appropriate to consider the industry perfection. This can be discussed in different sectors. Those belonging to the healthcare feel that a lack of understanding industry personnel have concerned long time cost and the electrical causes behind them. In the Textiles and Plastic industries, electricity is a small fraction of the production costs of most of the manufacturers — labour and raw materials comprise the largest percentage. For this reason, it is difficult to allocate funds for power related problems. In the hospitality sector, power quality generally receives low marks in priority compared to other power issues, such as opportunities to reduce energy uses and costs. Facility maintenance, customer service and personnel management fare much higher on the scale of importance.

IT intervention has remained important element in the overall systems and network management, basically for condition monitoring. The larger industries and for that matter even the utilities will like to switch over to what can be termed as e-power. The power quality and power availability along with functioning of all the equipments in any of the industry or the supply change could be known through the web. This concept is nothing new and is already being applied to a large scale in some of the operations in the energy sector; and to greater extent even in the manufacturing sector.

Improvement in distribution network will continue to pay an important role for the Indian industry to receive quality and reliable power that makes them globally competitive. Such a move forwards no doubt needs government's policies on one-hand, but it also needs a large number of other supporting technological and implementation strategies. The short-term and the long-term strategies will emerge from what goes into the Electricity Bill and the approach to the Five Year Plan. Recent advances of focus on control and instrumentation from the IT sector is welcome sign.

City of Delhi has been adversely affected due to availability of poor quality of power to its users, both industrial as well as domestic. With the service industries increasing their contribution to GDP growth, it is all the more necessary that quality and reliable power is ensured through reliable distribution network.

4

Impact of Energy Efficiency in Power Sector

Introduction

The World Energy Council has identified that the new modern technologies account for about 25 per cent of the potential improvement in power plant performance, with 75 per cent coming from better management and operational decision making. If the substantial gap between worldwide average performance and the top performing plants could be eliminated through the application of best practices, this would result in an estimated saving of up to US $ 80 billion per annum in building and operating capacity and a reduction in carbon dioxide emission of 1 GT per annum, as well as a reduction in other pollutants. In energy end use, prices influence electricity consumption pattern but cannot explain all the differences in performance among different countries. Completion and trade induce progress in efficiency of end use appliances. Framework conditions are improving and could foster efficiency gain of 20-30 per cent.

Accelerated technology diffusion and market reform measures are helping to reduce local and regional pollution from the production, distribution and use of energy services. Keeping energy options open so that nuclear power, hydroelectric and other renewables maintain or increase their share of the global energy mix is the best way to address global environmental goals in the short to medium term. The global governance of greenhouse gas emissions reduction should be pursued by governments, in consultation with industry, so that clear rules for emissions trading and clean development mechanisms can come into play as quickly as possible without upsetting national economies or

---

Paper presented at National Seminar on "Power Sector Reforms with a focus on Accelerated Power Development and Reforms Programme (APDRP)", during August 30-31, 2002 organised by CBIP.

excluding any energy option. Clean Development Mechanisms should be part of energy planning so that investments in new energy projects in developing countries which link commercial energy access with emissions reductions can be increased. It is important to acknowledge that the effective mitigation of global warming is the responsibility of all citizens as well as energy companies and government.

Initiative of the G-8

The energy ministers of the G-8 nations and top officials of many global energy companies met in Detroit during May 3-4, 2002 to define the roles that government and industry must play if consumers everywhere are to continue enjoying access to adequate supplies of affordable and clean electricity, coal, oil, and gas in the future. The outcome of the conference has relevance to the power sector foreign direct investments and growth in India.

G-8 nations — US, Canada, Italy, Germany, France, Britain, Japan, and Russia — which generate 70 per cent of world's economic activity and account for 60 per cent of its total energy use are now subjected to enormous energy related challenges. Specifically, the G-8 nations produce and consume 72 per cent of world's nuclear energy, 56 per cent of its electricity, 43 per cent of its coal, and 68 per cent of its natural gas.

The G-8 group considered challenges in three different areas as follows:

(i) Rapidly increasing demand,
(ii) Inadequate infrastructure to support future needs,
(iii) The need to reconcile energy growth with environmental protection.

The first issue refers to demand. G-8 nations will consume 33 per cent more energy in 2020 than they do today, even accounting for current and anticipated conservation efforts and the trend towards declining energy intensity in these countries. However, there is more to the world than the G-8; and the global demand for oil is expected to increase by an astounding 50 per cent in 2020 to 120 million barrels per day – half of that earmarked for transportation.

The second issue relates to infrastructure and the move towards tomorrow's energy does not exist today. The US Secretary for Energy cited two US examples to illustrate the extent of the challenge. Gas first. The US has a pipeline transmission system capable of handling of about 23 trillion cubic ft. of natural gas on an annual basis, which amounts to about 15 per cent spare capacity—today. In 2020, the system will have to handle 3030 trillion cubic feet. In Europe demand growth is similar but the challenge is different. Their growth requires infrastructure to move gas from Russia, Algeria and the Caspian region to where it is used.

According to the data of the US the electricity infrastructure challenge is even more daunting. The international energy agency estimates, that investment requirements for electric power plants and transmission lines worldwide, over the next two decades, will approach $ 2 trillion. Despite the recent downturn in the new power plant business in the US, the Energy Secretary is holding fast to its projection of a 45 per cent increase in generating capacity by 2020 — the figure that translates to the need for one new power plant weekly over the next 18 years. Plus associated transmission lines. Europe is similarly challenged, all order having to build the equivalent of its entire current electricity generating system by 2020.

The third issue of challenge is ensuring a secure energy supply accomplished in an environmentally sound manner. Striking the right balance between energy security and environmental performance calls for aiming to make the environmental cost negligible and the energy for growth benefit significant.

One of the goals of the G-8 meeting was identified broad areas where the member nations working together would accelerate progress in forthcoming energy challenges and ensuring future economic growth and prosperity.

It is significant to note the emphasis on 'energy security' by the G-8 countries: First and foremost on Abraham's agenda is to place greater emphasis on energy security and, in particular, on improved emergency preparedness to ensure adequate supplies in the event of disruptions. Second: open up and expand energy trade and investment. Third: increase cooperation in technology R&D.

An assessment of the goal of expanding trade and investment, the G-8 acknowledged that significant investments would be needed in energy development, production and infrastructure, as well as in improved energy efficiency to meet demand growth in an environmentally sound fashion. The group also felt that continued research, development, demonstration and deployment of a broad array of energy technology options will play an essential role in diversifying the energy mix and reducing the environmental impacts of energy production and use, thereby making a vital contribution to sustainable development. In line with G-8's goal of making clean, reliable and affordable energy available to all the world's citizens the ministers' agreed to improve cooperation in the development and commercialisation of energy technologies — particularly in the areas of emerging renewable energy, energy efficiency, and cleaner energy technologies.

Thrust on Efficiency in Power Sector by G-8

This consultation held in May 2002, virtually lays the ground rules for future development and investments in the energy sector in general and the power sector in particular. A message goes out clearly that for energy developments, from now to 2020 wherever new infrastructures are to be created investments will be difficult to access, even in the developed world. Wherever excess infrastructure capacities are available or there is possibility of efficiency improvement, the power generation can be improved. The end use efficiency improvement optimises use of supplied power. Thus, the developed world at this stage focuses on lower direct investment possibilities. That also indicates that investments in G-8 countries have many competing opportunities. If that be so then how can India hope for larger FDI in power sector Efficiency improvement route therefore becomes important for India.

WEC's Projection on Efficiency Improvement

The World Energy Council after its 18th World Energy Congress at Buenos Aires in September 2001, focused on internationalisation of energy services. The WEC reflected its unique position to sow the seeds of working together on a global, multi-energy basis. The recommendations called for pursuing broader and deeper best practices and efficiency improvement in energy production, distribution and utilisation, with a special focus on cleaner technologies and more rapid diffusion of technology to developing countries. Education on the role of energy in sustainable development, capacity building in developing countries, and better communication with the general public are some of the other important factors in this work.

The efficiency improvement over the next 20 years has been

reflected to have an average improvement of about 1 per cent per annum.

India's Thrust on Modernisation of Power Plants

In the Indian Power Sector scenario, most of the attention was focused only on the power sector reforms in the post economic reforms era. This focus tends to give an impression as if the process of modernisation and maximisation of output from power plants in India was only undertaken as a follow up of the 1992 reforms. Electricity is accepted by the Government of India as a basic minimum service, and action plan has been prepared by the Ministry of Power for the provision of electricity to all villages by the end of 10th Plan and all households by the end of the 11th Five Year Plan as per vision 2010: Power for All. It has a special component of formulation of national perspective plan 2011-12 for renovation and modernisation and life extension works of thermal power plants. The Government of India set up working groups from time to time to review the procedure of planned maintenance of thermal power stations. Two earlier reviews were done in 1975 and 1985. However, the renovation and modernisation of the existing power stations has been considered as the most cost effective option to achieve additional generation by virtue of the short gestation period and low cost. The Government of India recognised the importance of R&M way back in 1984-85, when Phase-I of R&M programme for 34 numbers of thermal power stations in the country was launched as a centrally sponsored scheme. During the 9th Five Year Plan, life extension programme on 28 thermal power units was in progress. In addition, R&M works on 150 thermal units were taken up which are at various stages of completion. Remarkable progress has been achieved.

In the year 2001 a working group was set up. After approval from CEA and the Ministry of Power, the Report had been circulated to all the state electricity boards/utilities during May 2002 for adopting recommendations made therein. Average reduction of plant maintenance period from 51 days to 28 days would be possible for 200/210 MW units; and from 63 days to 28 days for 500 MW units per year during the 5/6 year rolling plan for maintenance.

The incentive award have also been introduced for thermal power stations in four groups on the basis of specific oil consumption during the preceding year. Similarly incentive award for reduction in auxiliary power consumption have been offered for various thermal power stations. Awards worth Rs. 15 crore were distributed on 12th June, 2002 to Thermal Power Stations in the country. This small move of cash incentive to various power stations has gone a long way in effective utilisation of optimum power production capacities of thermal power plants as well as for maximisation of output. If we read this effort in the context of the meeting of the G-8 energy ministers earlier in the year it would be observed as appropriate that measures for supply side efficiency improvement will become a major input for higher energy production, at the lowest cost in the country.

India could attain six per cent GDP growth rate during the 8th Five Year Plan (1992-1997), even when a minimal addition to the power capacity only due to efficiency improvements resulting in power generation equivalent to 2,700 MW on the supply side and the saving of equivalent of 2300 MW on the end-use side. This additionality of 5000 MW contributed significantly to the power generation capacities in our country.

Let me make it clear that energy efficiency improvement alone cannot help us to overcome all our problems of power generation. Additional capacities of power generation will have to be installed. The efficiency improvement will only help us in getting the optimum or the maximum from the existing power stations and thereby optimising use of the infrastructure available in the country.

5

Technology: The Real Challenge of Energy Sector

Introduction

Global projections confirm that there are abundant resources in every region of the world to meet growing global demand well into the 21st Century. Arithmetic can be very elusive or discomforting. The arithmetic averages of the economists in terms of energy resource availability per capita and arithmetic totals of the reserves of the geoscientists can cause havoc for a large number of countries. Most of the developing countries, India no exception, suffer from the combined detrimental effect of the two, which is not simple arithmetic total but moves in multiples. It is essential for all regions and countries to diversify their energy portfolios by keeping all energy options open.

A new power generation industry is rapidly emerging to face the broad and deep changes occurring in the way energy business is now conducted and to meet the requirements for competitiveness and environmental accountability. The traditional power plant is progressively turning into a more complex operation, trading not only energy and capacity as commodities but also green credits and other services. If that be so then the renewables have an edge even in this sector. The dire need of the country is to develop atleast 20,000 MW from renewable sources of energy.

Role of Governments

Governments have a legitimate and essential role in energy policy and shaping regulation, especially as market reform leads to regional or

---

Welcome Address at the Celebration of International Sun Day by the Institution of Engineers (India), Delhi State Centre on May 3, 2002 at New Delhi.

even global energy solutions. It is important for governments to maintain their policy focus on market reform even in the face of economic difficulties. Skilled people, transparency and the rule of law are critical issues on which government policies must be well conceived and put in place. In the context of sustainable development, they should renew their commitment to fund adequate basic energy research, as well as the development and demonstration of advanced technologies (for example, on carbon sequestration, renewables, and the potential for coal bed methane and hydrogen). This needs to be done in the context of an inter-governmental agenda aimed at reliable, affordable, and sustainable energy services for all people in the world with the following specific goals:

- Bringing commercial energy services to the one-third of the world's people who do not now have access and by involving developing countries in the planning of this RD&D programme;
- Protecting the local, regional and global environment;
- Pursuing a long term focus which takes into account the inertia of energy infrastructure and the impact of competition on the horizons of industry; and
- Diversifying the portfolio of robust technologies, which are the only true hedge against the uncertainties of the future.

Strategic Energy Challenges

Four key challenges to enhance energy security are identified as:

- Commercial energy access for the two billion people in the world who do not have it;
- Political and legal stability at the global and regional level;
- The need to increase efficiency through competition and technology diffusion.

Two of the above mentioned challenges are of direct consequence for today's discussion:

Energy Efficiency

Our present course of increasing energy demand is pretty well fixed for the next 20 years due to expected global economic growth and the long-range nature of turning over existing energy infrastructure. Increased energy efficiency is penetrating that infrastructure at the rate of about 0.5 per cent per year, contrasted to 1.4 per cent desired in the Liveable World Projection of the WEC.

That does not necessarily mean that demand has to grow further after 2020 to support economic growth, but it does highlight a significant challenge. We will have to more than triple the rate of penetration of energy efficiency, meaning we will have to find ways to turn over existing supply and end use infrastructure much faster at the same time that we avoid technological barriers to ever increasing energy efficiency. Some economists believe that penetration greater than 1 per cent per year is not possible, so this is a challenge we need to understand better. The pragmatic approach will be to effect efficiency improvement but also try for renewable and other sources of energy.

Use of all Sources of Energy

Nuclear Power

Nuclear power, large hydro and greenhouse gas emissions: For base load electricity generation, the most effective means currently in use to reduce CO_2 emissions are nuclear power and hydroelectric power. Those countries with the highest protection of nuclear and/or hydro power have the lowest CO_2 emissions per KWh. Nuclear energy and large hydro advantages in terms of global warming, cost stability and high capacity factors which make them compatible with the goals of sustainable development for tomorrow's world.

Renewables and distributed generation based on local resources: The time frame for substantial penetration of new renewable technologies in the global energy mix is now 3-40 years. In the short term some renewables, such as wind and geothermal, are a very suitable complement rather than a replacement for large scale generation. The space of development and use of renewables and distributed generation should be accelerate through increased RD&D and government support of development, especially in developing countries.

Renewable Energy

The Liveable World scenario counts heavily on renewable energy. The challenge of renewable energy is that it tends to involve what we call low "energy density". Think of energy density as the drive available in the form of pressure, temperature or velocity to move the machinery and produce useful energy. For example, conventional machinery for electricity production operates at extremely high pressure and temperature, which allows the machinery to be relatively small and inexpensive per unit of output.

In contrast, renewable technologies such as wind, wave/tidal and ocean thermal power must extract energy at low energy density, making the machinery very large and expensive per unit of output. Not enough attention has been given to this physical obstacle; as a result projections of future contributions by new renewable energy have been wildly optimistic for nearly 30 years.

Solar energy also begins with a low energy density, but sunlight can be focused at high energy density through the use of technology, which has been known and available for hundreds of years. The initial research on solar energy recognised this advantage, but when it was found there were commercial applications for low energy density photovoltaic applications not sensitive to high cost, attention shifted to photovoltaic; the research on concentrated solar energy was set aside. We need to get back to research on focused solar energy; otherwise, sourcing 35 per cent of the world's energy needs by 2050 from renewable energy is probably just a pipe dream.

Alternative Fuels

Hydrogen and methanol are mentioned in the Liveable World scenario. We must recognise that neither is a primary energy resource and must be derived somehow from fossil, renewable or nuclear fuel. Today, another alcohol, ethanol, which is produced in the central United States and Brazil as an oxygenating additive for gasoline, contains less energy than the energy required to produce it in the first place. Hydrogen produced by electrolysis has the same disadvantage. The pollution from energy use can be moved around, but cannot always be eliminated by the use of these alternative fuels and the amount of primary energy required must often be increased. At the same time, moving the pollution from a populated to an unpopulated area is, sometimes, worthwhile so ethanol is currently used to attack local air quality issues. We must increase our understanding of these alternative fuels, how they might be better produced and how they might ultimately contribute.

Renewables based systems in developed countries are a matter of add-on and luxury. The efficiency of operation and availability are not a concern for the user. In India, renewable based systems are a need and every equipment has to operate reliability during its life time.

The number of unelectrified villages will go up significantly, with the change in definition of electrified village from 'one-bulb electrification' to at least 30-40 per cent house electrification.

Conclusion

The real challenge in the energy sector is of Technology – from basic R&D to Product Development — and not Finance for projects. India has to focus both on base-load centralised systems as well as small-decentralised systems.

It is well known that sustainable energy supply is the combination of three aspects, namely supply security, economy and environmental protection. In addition to this, power station investments with running periods of 25 years and more require long-term planning security, which must not be jeopardised by political representatives out of periodically occurring tactical considerations.

We have to commit to seek energy approaches, which will take the world along the path of sustainable development. Our premise is that sustainable development includes the goal of creating the open economic conditions, which will continue to allow nations to raise themselves up out of poverty. The absence of poverty could well be the best indicator of sustainability. In this year of the World Conference on Sustainable Development the biggest contribution of energy experts will be to utilise the environment friendly renewable energy sources.

6

Research and Development in Power Sector: A Vision

Introduction

The power sector scenario over the next 15 years is projected to move in a highly progressive mode. Projections from the Central Electricity Authority for year 2011-12 and 2025-26 are highly demanding. This virtually indicates that India has to add power generation capacity at a rate of over 10,000 MW per year. The target is projected to be achievable if foreign and private investment is available. Although this is country's need-based requirement but is not supported by market forces. The observation over the past 10 years, since the invent of privatisation and liberalisation, show dismal results and do not lead to encouraging signs of high private investments at the level the country needs.

Power Sector and Private Investment

Two basic elements that need to be set right for investment and finance are:

(i) pricing of electricity; and
(ii) technological capability of the country.

Pricing of Electricity

It should be noticed that in the past, in many countries, the energy

---

Paper presented at National Conference on "R&D in Power Sector — Identification of the Needs and Vision Formulation for 15 years" organised by Central Power Research Institute during May 24-25, 2001 at New Delhi.

sector has been publically owned. Also in most developing countries, substantial international funding has supplemented the limited domestic capabilities. The share of private sector capital has usually been less than 20 per cent of total sectoral investment during 1974 to 1995. More recently, the growing private and public debts in industrialised and developing countries alike has made the energy sector finance, with its long amortisation periods, more difficult. Privatisation, however, is becoming an accepted political remedy. Same is the situation in India. But, for it to be success, it is necessary that the pricing structure reflect real costs.

Technology Development — The Key to Power Sector Growth

The other element is technological progress. Innovation, especially commercialisation of novel technologies and processes, require continual innovation of efficiency in R&D. In turn, technology diffusion depends on both RD&D and steady improvements through 'learning by doing' and 'learning by using'.

Innovation in technology diffusion requires that opportunities are perceived and that entrepreneurial spirit exists to pursue them. Long term scenario cannot forecast future technological 'winners', but can indicate the *areas of technological opportunities*. This is what can be projected for 2015.

The present global scenario indicates that the technological scene from now to the year 2015 can be visualised on the basis of the technologies already identified or which are at the advanced stage of development.

Characteristics of RD&D

There are a few characteristics to be noted:

(i) *Technological progress has a price* — continual investment in RD&D is necessary. All the technological improvements need a steady flow of RD&D investment. There is no place for stop and go strategy. RD&D in clean and efficient energy technology and infrastructure is the real strategy for future.

(ii) *RD&D expenditures are declining* as the government thought that advanced technology will flow along with foreign investments. Neither took place. Private investment in RD&D is declining along with public efforts in RD&D. This expenditure is increasingly viewed as too expensive in the market where maximising short term value takes precedence over long term environment protection.

(iii) Consistent *emphasis on clean forms of final energy* suggests that in the long term, meeting even high levels of end use energy demands can be consistent with regional and global environmental protection. For near term, it is important to accelerate the diffusion of existing environmental technologies. But for the benefit of long term, it is important that near term solution do not lock in the existing technology and slow the introduction of cleaner fossil and non-fossil alternatives.

(iv) Technological advancements to improve and expand infrastructure for clean grid-dependent energy forms is crucial for industrial growth, but at the same time, *the distributed power has to be considered for reaching energy at an affordable cost for the poor* in remote areas.

(v) *The traditional form of energy technology policy has focused on energy supplies*. But, WEC has projected that long term improvement in energy system cost, efficiency and environmental impacts is increasingly determined by technological innovation at the level of energy end use. Over the long term, much of today's tradition may shift towards the end user.

Technology Issues to be Tackled

The technology issues that need to be discussed include the following:

(i) Identify technical problems for supplying electricity at affordable cost to the rural poor and develop implementation strategy;

(ii) Improve efficiency in the existing power plants through retrofit to ensure their operation at a high plant load factor and high availability factor;

(iii) Introduce state-of-the-art technological intervention to bring down cost of power generation;

(iv) Ensure supply of quality power;

(v) Conduct comparative analysis of Distributed Power (defining their scale of operation, say up to 25 MW) with large market and centralised power with limited market.

(vi) Accelerate development of new energy technologies, particularly the next generation fossil fuel technologies that produce 'near zero harmful emissions';

(vii) Promote higher use of nuclear power on the basis of the established record during the last few years;

(viii) Ensure increased use of hydro and renewable energy based power generation;

(ix) Focus on more efficient use of energy, especially at the point of end use in buildings, electric appliances, vehicles, and production processes;

(x) Promote new enabling technologies for dispatch, monitoring and control of power;

(xi) Identify technologies for effective distribution and transmission to bring down losses; and develop implementation plan;

(xii) Establish a benchmark for transmission and distribution losses on an All India basis as well as district basis (and not to be carried away by the benchmark followed in the developed countries);

(xiii) Identify and develop technological solutions for accessing power from neighbouring countries and its transmission through the grid (promoting effective regional trade in power).

7

Nuclear Power Development

The Energy resources round the globe confirm their availability in abundance in every region of the world to meet the growing global energy demand well into the 21st Century. It is essential for all regions and countries to diversify their energy portfolios by keeping all energy options open, though the fossil fuels are expected to remain the dominant source of the total primary energy mix for several decades.

A new power generation industry is rapidly emerging to face the broad and deep changes occurring in the way the energy business is now conducted and to meet the requirements for competitiveness and environmental accountability. The traditional power plant is progressively turning into a more complex operation, trading not only energy and capacity as commodities but also green credits and other services.

The key linkages addressed at the 18th Congress of the World Energy Council in October 2001 in Buenos Aires focused on following issues:

(i) Oil and gas price volatility and decoupling,
(ii) Cleaner combustion technologies for oil, natural gas and coal,
(iii) Gas/electricity convergence and multi-energy services,
(iv) Gas and potable water,
(v) Nuclear power, large hydro and greenhouse gas emissions,
(vi) Renewables and distributed generation based on local resources,
(vii) Competition and efficiency.

Use of nuclear power was under cloud after the Chernobyl accident.

Chariman's Remarks at the Energy Summit (Session on Nuclear Energy) organised by Confederation of Indian Industry on February 6, 2002 at New Delhi.

Its role in meeting base load needs has been now realised for environmental reasons, leading to its revival worldwide. Those countries with the highest proportions of nuclear and hydro power have the lowest carbon dioxide emissions per KWh. Nuclear power plant have advantage in terms of restricting global warming, cost stability and high capacity factors which make them compatible with the goal of sustainable development for tomorrow's world.

In case of nuclear power there is a range of issues including plant life extensions, new plants, reprocessing of spent fuel to maximise its use (where economically feasible), and innovative technologies which address design, licensing, fabrication, construction, performance, safety and effective waste management. The nuclear industry recognises, however, that with the help of government it must continue to improve the public acceptance of this significant source of energy. Nuclear Power Corporation has focused its attention on all the issues that were raised against the use of nuclear power for the benefit of mankind.

I remember a Society for Sane Nuclear Policy was set up in 1977. Lot of hue and cry was raised against nuclear power use in India. However, consistent efforts of the Nuclear Power Corporation starting with its focus on re-conditioning and rehabilitating the Rajasthan Atomic Power Plant Unit I and II in record time; at low cost utilising indigenous know-how. The documentary evidence of the same and awareness in public of the achievement could mould opinion in favour of nuclear power. Continuous operations of power plants and highest environmental protection measures has brought public interest back into nuclear power in India.

This indicate the nuclear power programme in India maintain highest standards of environmental management; supports nuclear power for base load applications.

I remember when in 1999 the Rajasthan Atomic Power Plant applied for National Golden Peacock Environment Management Award, instituted by the World Environment Foundation, the irrefutable data provided by them was the basis for their being awarded in the category of large units. That was the first time that any nuclear power plant in the world was awarded for environmental management. The Award was presented at the First World Congress on Environment Management in June 1999. Since the European Union's environment programme and the European Partners for Environment were also actively involved in that conference, a large number of foreign participants vehemently opposed the award to a nuclear power plant.

The scientists and technicians from nuclear power plants produced video and other documentary evidence to prove the performance of the power plant. That did bring about change in the participants who before leaving the conference appreciated that nuclear power if properly handled can be made environmentally sound. And India has shown the way in the subsequent years, the Atomic Power Plant at Narora, and the Heavy Water Plant at Managuru also received Golden Peacock Environment Management Award.

The Perspective Planning Group of the Planning Commission preparing a Development Plan for 2020 also agreed on the possibility of nuclear power contribution up to a level of 20,000 MW out of a total of about 320,000 MW required by 2020. Recent advancement by Nuclear Power Corporation are encouraging and if current estimates come through then the Nuclear Power Corporation could even achieve higher generation capacity target.

As the Conference is focusing on the industry, let me point out that there are three basic requirements for interaction. The industry needs to *reduce cost, increase efficiency and respect the environment.* It appears the current nuclear power programme has all the three elements inbuilt. The government can help the vendor industry by complementing private RD&D investments with support for basic research and demonstration of new technologies, by providing captive market and financial assistance. It is important for the government to maintain its policy focus on market reforms even in the face of economic difficulties.

Skilled *manpower, transparency* and *conformance to rules* are the critical issues on which good corporate governance relies. In case of nuclear power programme all these three pillars have a sound footing.

Continuous increase in internal revenue generation has projected NPCIL as a *bankable* organisation.

Nuclear power option for base load is logically sound; in the interest of— energy security, national development and common man's prosperity.

Reaching energy to the poor through nuclear power generation has to be ensured keeping in focus Dr. Bhabha's words of wisdom 'No Power' is costlier than 'No Power'. Recent advancements have made nuclear power supply cheaper than competitive thermal power supply.

8

CNG for Transport

CNG and the bottlenecks in its supply faced by the citizens of Delhi, bring forth an important issue of evolving energy solutions to the challenge of securing energy supply and addressing it with due awareness of energy use and the relationship of both to broader social and environment aspects of sustainability. Unless they are set in a wider, more holistic context, they may create other problems and fail to bring the expected benefits. There have been positive contributions that human uses of various energy sources have made to the lifestyles and economic and social development of many people. But a lot of people have not benefited from these positive contributions — and many have suffered negative effect on their health and welfare. To varying degrees, they in turn reflect a loss of bio-diversity and species reduction and over-exploitation of land and water resources. CNG supply in the city of Delhi has brought forth the issue of energy use on health. Though it does not directly reflect the non-availability of energy resources to large population.

There is serious doubt whether the present trajectory of human economic development and social organisation can maintain the quality of our human life and the natural world around us. Without due care and constraint, there is a threat of an unliveable world. Efforts to bridge the apparent contradiction between economic development and protection of the natural environment for both the present and future generations, focus in part on a minimum or fixed talk of each main component of natural capital. Under this approach any development path which leads to a reduction of natural capital below a minimum of fix level is defined as unsustainable, even if other sorts of capital increase. Civilisations have collapsed because they fail to remain in tune with their environment. Thus the focus on energy use has to be on sustainability. Despite the

---

Address at the Workshop on Use of CNG organised by Jamia Millia Islamia on April 30, 2002 at New Delhi.

conceptual contradictions of sustainable development and difficulties in putting it into practice, it is not difficult to understand the desirability of meeting the needs of the present without compromising the ability of future generations to meet their own needs. There is a grate deal of wastages in energy use and the scope for increasing efficiency and reducing waste in end uses is also considered. Every inefficient use is linked to higher pollution. Therefore, a focus on optimum efficiency of operation and sustainability of systems is extremely important.

The concept of sustainability was brought to centre stage by the shattering of 3,500 Sqm Larsel B Iceshelf in Antarctica, which proves once again that human activity has a colossal impact on the environment. Humans as a species cannot survive unless we so arrange our activities and processes that we practice environment friendly processes. Certain sustainability indices include strategy, innovation, governance, stakeholders and health.

Use of energy for development whether economic or social has the direct impact on sustainability. Certain data report that the coal based power stations contribute to about 45 per cent of power generation, but roughly 80 per cent pollutions from power generation. On the other hand, the gas based power generation contributes to 24 per cent of power generated on the globe and only 17 per cent of power related pollution. In this context, use of energy in any form and specially the health impact of use of such energy sources have to be viewed. Use of CNG is also to be viewed in the same light.

The Supreme Court ruling of April 5, 2002 is momentous. It moves away from the beaten track of issuing deadlines, expecting the government to think through and implement the CNG order. This time the three-Judge benches headed by Justice B.N. Kirpal has simply detailed out an implementation strategy for the government to act upon. The ruling hits hard at the government's incompetence, even to perform its administrative role, by reminding it of its fundamental duty to protect right to life and thus people's health.

This ruling came after extensive hearing on all niggling issues, including availability, supply and allocation of CNG, issues of ultra low sulphur diesel and doubts about CNG technology being experimental. The direction is comprehensive and conducive. It is expected to put the CNG controversy at rest unless the BJP led Central Government once again tries to be subversive as is evident from the ongoing move to kill it with an ordinance. Fortunately, the Congress led state government has been more responsible in its reaction and action. Both the governments will have to get their act together and implement the order.

Slapping a fine on the Union Government for repeatedly seeking review of the order and thus abusing the process of the Court, has signalled a strong message. The government will have to drop its saboteur stance. At a time when both the central and state governments are out to stymie the small CNG market with harsher taxes, the Court has shown where the fiscal penalty should be imposed – on dirty diesel and promoters of dirty fuel. This is the first instance ever in the country of a fine being imposed on a polluting technology based on the polluter pays principle.

This time the entire nation stands to gain from this ruling thanks to the devious attempt of the government to scuttle the CNG strategy by tabling the interim report of the A.R. Mashelkar Committee, to allow the use of low sulphur diesel. The Supreme Court calls the bluff. Rejects the plea. But uses the opportunity to broaden the ambit of this case beyond Delhi to give a direction on alternative fuel strategy for the whole country.

This notable ruling that upholds the supremacy of human health must not be diluted only to protect the economic interest of a few thousand-diesel bus owners that are holding the city of ransom.

Report comes under fire

Mashelkar Panel's suggestions do not address issue adequately, observes court

The Supreme Court (SC), in its April 5, 2002, verdict has frowned upon the recommendations made by the Mashelkar Committee. "The report does not show any serious concern in protecting the health of the people." the SC noted.

The panel was set up in 2001 by the Union Government to submit a report on vehicular pollution. It was headed by Council of Scientific and Industrial Research (CSIR) Director General R.A. Mashelkar. When it gave an interim document on its findings in January 2002, the Union Government lost no time in accepting it as the final report. This despite the fact that it was meant to be debated by the public. The reason for the undue haste: the Committee recommended a multiple fuel policy for Delhi, which also included diesel.

The SC has, however, rejected the government's acceptance of the report describing it as a means to contest CNG's (compressed natural gas) credibility as a clean fuel.

The Mashelkar panel had suggested that while emission norms

should be declared, the choice of fuel should lie with the users. Interestingly, India already has such norms but their implementation leaves a lot to be desired. "It is naïve of the Mashelkar Committee to expect that merely laying down fresh emission norms will be effective or sufficient to check or control vehicular pollution," the Court stated.

While broadening the jurisdiction of the case to other polluted cities beyond Delhi, the Court said that the Mashelkar report would extend the suffering of the people of these cities from air pollution. The fact that the Mashelkar Committee did not have any medical expert as its member also drew flak from the Apex Court.

It is a pity that the Supreme Court had to involve in executive and administrative strategies and issue a detailed implementation strategy. The general impression is that the automobile manufacturers and users, mainly the commercial transporters, were willing to adhere to pollution norms at an early date and the Dr. Mashelkar Committee has served their interest at the cost of public health. It is basically the public perception that count since there are two committees, one set up by the Supreme Court and the other by the Government of India and both them are headed by an official at the level of the Secretary to the Government of India have equal and similar mandate and intellectual acumen. The bona fides of neither of them could be doubted. Even a number of members of two committees are common. But the outcome is so diverse. One could only say that Dr. Mashelkar's Committee was supporting the government's viewpoint, whereas the Mr. Bhure Lal Committee has reflected the citizens' viewpoint or what will be held right under the interpretation of law.

This is a serious issue and has certainly raised many other concerns as to whether we should put the common man at the centre of the activity while taking decisions that have direct bearing on the health and general being of the citizen or else we should just focus on government's capabilities to handle the situation.

The problem with CNG is specifically of technology and pricing. Non-availability of appropriate dispensing mechanism reflected the hollowness of technical capabilities of the concerned agencies. In case of CNG supply, it is the compression of the gas at the last supply and that is crucial. And when a problem comes with this system, it has directly and adversely affected the common man. Like in every other energy sector, the technology comes first and anything later. The widening gap of technology divide has to be bridged, whether it is the manufacturer of the equipment dispensing of the fuel or anything else.

9

Rural Energy: Role of Electricity

Introduction

Global energy policy makers have realised the poor living condition of over 2 billion people in the developing world, who have remained isolated from electricity and commercial sources of energy. Are these communities really deprived of all kinds of energy sources and the socio-economic development opportunities? Closer scrutiny shows that the rural poor, in the developing countries without electricity and commercial sources of energy, have the means to meet their minimum sustenance energy needs in an appropriate fashion through biomass use. The World Bank projected (in 1992) a figure of 2 billion who were undergoing commercial energy poverty. The global advice was to develop market based energy supply systems and ensure accessibility to such markets by the rural poor. The World Energy Council in its report titled "Global Energy Perspectives" (1998) projected that such a market mechanism was useful only for those who have a role in market economy. These 2 billion are poor and have no purchasing capacity to take advantage of the market based system. They are a case of market exclusion. All such population survive on biomass fuel, which under a number of situations in the developing world are available free of cost. The biofuels are, therefore, termed the poorman's fuel. The accessibility to commercial sources of energy through market mechanism has not become a reality.

It is calculated that the amount of energy, roughly equivalent of 7 per cent of the world's electricity production today, could cover basic human needs. Demographic trends are taking serious turn. Most of

---

Based on WEC Report: The Challenge of Rural Energy Poverty in Developing Countries, 1999. The Author was a member of the WEC Rural Energy Steering Committee.

today's 2 billion people without adequate energy services are in rural areas, but urban populations are projected to grow more rapidly. Far from easing rural energy poverty, this is likely to reinforce policy makers' occupation with urban issues while increasing competition in rural energy supplies. Disease, in particular AIDS will also take its toll on the development capacity of the worst affected countries.

Rural Electrification

Most investments in the energy sector of developing countries are targeted to modern energy sector, though relatively modest investments in now dominating traditional energy sources can potentially make a significant difference. In keeping with the bias to the modern energy, and electricity is key to the modern world, most developing countries have rural electrification programmes. Sometimes these promote decentralised electrification, for example with rural energy sources, moreover, the focus is on grid extension. Although the number of rural households with access to electricity doubled in the 1970-1990 period, this barely kept pace with population increase.

In general, the costs of electrification were underestimated while its benefits were overstated. By itself, electrification does not guarantee economic development and its benefits tend to accrue to the wealthier groups in electrified areas. It is increasingly recognised that electrification must be part of much broader development package. Electrification contributes to, but is not a substitute for other interventions.

WEC Report on "The Challenge of Rural Energy Poverty in Developing Countries" carried out a review of rural electrification with a view to give a perspective on present situation. It has been observed that the global total mark the significant variation between region and countries, for example, a number of Asian countries have made impressive progress, most notably China, Malaysia, Thailand and Philippines. China's rapid electrification programme in the period 1970-90, through which 365 million more rural residents gained access to electricity, significantly increases world total.

While the traditional sector needs higher policy priority, rural electrification is a legitimate development objective, indeed, development beyond a certain level is virtually impossible without electricity. In addition with technical progress the gap between the electrified and the non-electrified world continues to widen. At the same time it must be stressed that the electrification outside the context of rural development initiatives normally has insignificant, if any impact.

The dilemma facing many developing countries is that they recognise that they required resources are in short supply. Rural electrification policy making is thus about making difficult choices about where, when and how best to carry it out. This demands clear thinking of (i) How the benefits of rural electrification are described? (ii) How much consumer should pay for obtaining them? (iii) What criteria should be used in the selection of areas for electrification? and (iv) Supply technologies.

Some Examples of Rural Electrification Programmes in Different Countries

Rural electrification programmes have typically concentrated on connecting villages and remote areas to a national grid often owned and operated by the public utility. The tendency is to extend the grid incrementally, reaching towns and settlements in order of increasing capital costs. This principle of economics has always resulted in remote areas with small populations to receive the electricity supply at last. By contrast the cost of installing decentralised generation technology is independent of the status of the transmission and distribution system. Thus, in many remote settlements it is common to find diesel generators either for a single user or as a part of local distribution network.

Some countries have pursued decentralised electrification as an important part of their rural electrification strategies. In China, there are about 60,000 small hydropower stations with a total installed capacity of more than 17 GW, supplying nearly 20 per cent of rural electrification consumption. In the Philippines and Indonesia, where a large proportion of the rural population lives on islands, decentralised options are often the only way to supply electricity, primarily through small hydropower systems or diesel generators.

Thailand is considered a success story and its grid-based rural electrification programme, which began in 1974, has increased the number of electrified villages from 20 per cent to a current 98 per cent. Its success is credited to careful expansion planning, efficient billing, a cross-subsidising rate structure designed to charge large users higher amounts than small users, responsiveness to customers and good marketing programmes.

Kenya's programme provides incentives to consumers to connect to the electricity grid where it is available. Under this programme, the consumer is required to pay only a token connection fee of approximately US$ 40 and a small meter charge of US$ 5. This should be compared to the actual cost to the utility, which are estimated to be over US$ 1500 per rural connection. Connections have been averaging 5000 per year during the 1990, but given the levels of subsidy, the programme is hampered by lack of funds. Only 4 per cent of the electricity supplied by the national grid reaches rural customers.

Rural electrification co-operatives present a viable public sector approach. In Bangladesh, the appalling financial and technical failures of the public power utilities led, in 1980, to the government supported take over of its semi-government organisations by rural electrification co-operatives based on US model. By 1996 their number had grown to 45. In 16 years they have engineered a rapid expansion of grid based rural electrification supplies, which now serves 1.6 million consumers, as many as the public sector in urban areas. The co-operatives have created and fostered an alternative culture and structure to meet a demand that was unexpected in such a poor country.

Argentina has recently begun implementing an innovative rural energy plan designed to encourage private sector involvement in providing rural energy services to 8 provinces. In each province private companies bid for the right to provide electricity to the 1.4 million people and 6000 schools, medical centres, drinking water facilities and other public services without access. Preliminary analysis show that renewable technologies will be competitive with diesel generators; in most cases a large share of household supply will be through solar photovoltaic home systems.

The Impact of Rural Electrification

Surveys have shown consistently that a number of factors often make rural electrification programmes more meaningful and successful by, for example, generating markets for electricity, leading to higher rates of return on investment. Rural electrification clearly contributes to, but it is not a substitute for, other rural development interventions. As simple and straightforward as this may seem on hindsight, it has often been overlooked in the formulation of rural electrification programmes.

Many earlier rural electrification interventions focused on extending the grid. Clearly electricity is more versatile and once installed can usually allow for most conceivable increases in demand. However, decentralised options may be attractive for a number of other reasons. Where the demand is uncertain or latent, a diesel generator, for example, will require a lower initial investment and hence the risk of substantial financial loss is reduced.

Major Issues in Renewable Energy Technology Use for Rural Electrification

There are three important issues to consider when looking to renewable energy technologies to solve rural energy problems. First, in some places the diffusion of renewables may be going too fast. Second, subsidies on many conventional energy sources, particularly diesel, kerosene and electricity for agricultural purposes, still seem to be significant, although coming increasingly into question. Third, while the potential market for modern rural renewable and other energy technologies is clearly enormous, the surface has hardly been scratched.

Access to Credit and Rural Energy Development

A key form of intervention in the bid to increase rural people's access to energy is financing. Some of the countries have taken bold steps of reducing subsidies on prices and making micro-credit availability in large scale which has helped in dissemination of improved energy services.

Some of the important interventions have emerged in Bangladesh by the Grameen Bank through its Grameen Shakti which aims at 100,000 photovoltaic systems to be operating in Bangladeshi rural homes by the end of 2000. Indonesia has been successful in selling solar photovoltaic home systems in the rural areas through innovative credit arrangements and services. By combining all of the operational and financial functions at the local level it was possible to open up new markets and also serve the needs of rural communities. The Dominican Republic and Honduras successfully demonstrated viable credit financial market for delivering household photovoltaic systems to non-electrified rural villages. A major company developed the concept of providing photovoltaic household energy on monthly leasing basis. IREDA is the only financial agency in India providing soft term loans for promotion of renewable energy systems. By mid-1998, IREDA had sanctioned 963 projects covering over US$ 300 million.

The total recovery rate has been as high as 99 per cent. SELCO is one of the solar energy services companies marketing small-scale photovoltaic power systems in southern India. In order to access funds for direct consumer financing, SELCO had to seek a bank guarantee from a US not-for-profit company. As a result of this guarantee, SELCO negotiated with IREDA to access World Bank Global Environmental Fund for on-lending to end-users.

Future Directions

The report has looked into the future directions in rural electrification that will really make the rural areas develop in the new information age. Following specific interventions have been identified:

(i) Technology for lower capital cost regional and local distribution, metering and billing systems.
(ii) Technology for sound connectivity of regional and local distribution networks with the grid.
(iii) Distribution technology for low capacity end users.
(iv) Metering and billing.
(v) Electrical energy service technologies.
(vi) Energy efficient lighting.

Conclusion

The Report provides an insight into the action taken in rural electrification so far and their impact. It also gives new directions to rural electricity supply and its linkage to rural development. It has very appropriately linked the reasons for poverty in rural areas with rural energy poverty. This is an objective study which reflects the situation in most of the countries and provides an opportunity for intelligent discussion and application.

10

Energy for Agriculture

Introduction

The efficient food security system projects a close relation between the quality and the quantity of food produced, transformed and consumed and the quality and quantity of energy used to power agricultural production. The situation in many areas is characterised by a lack of sufficient energy inputs to satisfy the minimum requirements of food security. Dependence on animal and human power is a reflection of the same. Food losses are high due to the lack of energy for processing or preparation of storage or for transportation to markets.

Status of Agriculture in India

Although considerable effort has been made in a number of developing countries to promote long-term energy sources for agricultural activities, most of such efforts have failed. Many reasons can be found for these failures, but often it is because they have not been introduced with enough sensitivity to rural socio-political or economic constraint, or they have not taken into account gender or land and resource tenure issues. The agricultural sector, if not entirely overlooked in national energy policies and programmes, is seldom given the priority it deserves. Planners in the energy sector seldom co-ordinate with those in the agricultural sector to ensure the improvement of agricultural activities through energy availability, by providing energy sources of the right from, cost and quantities, in rural areas.

Paper presented at the National Workshop on Energy and Environment Management for Sustainable Development of Agriculture and Ago-Industrial Sector organised by the Institution of Engineers (India), Madhya Pradesh State Centre, during July 8-9, 2001 at Bhopal.

Traditionally, the agriculture sector has relied on non-commercial sources of energy — mainly the draught animal power and the manpower. Low energy input into agriculture has often affected sustained employment and income generation for a large rural population. The NSSO data for 1995-99 indicates that inspite of GDP growth in other sectors, the agricultural and rural sector continue to provide 62 per cent labour employment in the country. There has been a debate on suitability of mechanised farming in a poor country like ours as that can mean poor uneducated labour going out of employment. On the other hand economies of scale demand that mechanised farming be adopted for cost effective agricultural production. Another observation during the last 5 years is that with the introduction of mechanised farming most of the women labour, which was earlier involved in harvesting, is now redundant, thereby making them unemployed.

No growth pattern in India can really be useful unless it leads to higher employment opportunities in the rural and agriculture sector.

Food production and socio-economic status of the people have been changing since independence. Agriculture is projected to continue to be the backbone of Indian economy, up to 2020 and beyond. Agriculture and rural development will also provide the largest labour employment opportunities. While the total food grain production rose from 51 mt in 1950-51 to over 200 mt this year. There is a concern for the current declining trend, which can only be arrested through better technology and higher inputs. It is estimated that for nutritional adequacy, the demand for total food grain will reach about 300 mt by 2020. The current practice of cereal production is also likely to diversify. The suggested route will be from low value crops to high value crops like fruits, vegetables, flowers and other horticulture crops. This, in turn, will depend on the generation of adequate surplus through higher investment in rural infrastructure and in scientific research including the broadening of research efforts from wheat and rice to include newer crops and allied agricultural activities. The availability of assured irrigation continues to be essential precondition for the adoption of new seed fertilizer technology in most of the crops. Both hybrid seed production as well as better irrigation *call for larger energy inputs*. Efforts are being made to devise appropriate institutional strategies to involve the small and marginal farmers so that they may also be able to avail of the benefits of increasing growth, agricultural diversification and of exports.

Gap between actual and potential in productivity is quite large. Paddy production and coarse cereals need to be produced at higher level to achieve sufficiency. The possibility of developing processing industry for extracting edible oils from non-oilseed commodities, like rice bran, is also being explored. The introduction of palm cultivation for oil production is expected to reduce pressure on traditional oil seed crops to meet future edible oil demand. In case of sugar cane research the development efforts are being strengthened to increase the yield per hectare by about 4 per cent per annum. The demand of sugar can also be met by developing mini sugar mills so that substantial sugar cane production can be diverted from khandsari to sugar production. These are also *energy consuming*.

Energy Use in Agriculture

Agricultural energy needs can be divided into direct energy needs, such as those required for land preparation, cultivation, irrigation, harvest, post-harvest processing, storage and the transportation of agricultural inputs and outputs; and indirect energy needs in the form of fertilizers, herbicides, pesticides and insecticides. Most of the energy discussion refer only to direct energy need and the same criteria is used in this paper.

Available data from Asia indicates that indirect energy inputs form, on an average, 64 per cent of the total energy consumed by the agricultural sector. The total includes both direct and indirect energy inputs, inclusive of commercial and traditional energy sources. This data does not include human or animal power since the conceptual and statistical validity of such an inclusion is debatable and it is almost always ignored in energy data compilations and analysis. That said, the bulk of energy inputs into land preparation, cultivation, post-harvest processing and agriculture-related transport in developing countries, particularly in the subsistence agriculture sector, is in the form of human or animal power.

In 1980 the FAO estimated the division of the use of agricultural power in developing countries as 5 per cent mechanical, 29 per cent animal draft power and 60 per cent human.

In view of the arguments given above details the paper skips on the human power and the animal power in agriculture. For reference it may be noted that a sustainable rate at which the human body consumes energy is about 250 watt to 300 watt for a fit person. His efficiency of energy conversion is only about 25 per cent, this means that the maximum sustainable power is only 75 watt. In case of animal power, power output in Watt range from 200 W for a donkey, to 650 W for a camel, and 520 W for a buffalo.

Machine Power for Production Agriculture

Low power tractors of 30-35 horsepower would ideally suit some of the large land holdings in India where a power take-off can operate rotary hoes, pumps (for irrigation and crop spraying) and threshing machines. In Asia the single-axle, two-wheeled tractor predominates in the irrigated rice and horticultural crop sectors. The cost of tractor operations are such that it requires about 40 ha of good arable land to support the costs of a 30 horsepower tractor. Such large tracts of land are generally not available with most of the farmers, thereby making it necessary for co-ownership and co-operative use of such tractors to achieve viability.

Additional energy available with machine as compared to animal and human power needs to be put to the best possible use. In short, the experience, judgment and skills of humans needs to be applied in a synergistic manner with machine power-intensive operations. In addition, the power source used should be user-friendly; economically viable: as environmentally sustainable as possible; adequate; easily available to the farmer; and add value to the time saved.

Despite the many advantages to be gained from it, the degree of agricultural mechanisation is generally low in India, though there has been a growth recently.

The inputs of commercial energy are increasing only slowly, except for irrigation.

Energy for Food Preservation

An important and often overlooked use of energy in the agricultural sector is in the preservation of foodstuffs, a vital link in the food security and hence rural welfare chain. In addition to processing in general, packaging and distribution also require energy. The food products available to today's consumers have an energy price tag and the relative cost of cooking, preserving and packaging need to be considered when developing food products. The extended shelf life and storage capability of modern foodstuffs is a source of food security particularly to urban communities at a distance from the farm. However, for greatest economies and greatest returns to the farmers, it is often best for these processes to be carried out in rural areas, which results in additional energy demand.

Many food crops when harvested cannot be consumed directly but must pass through several stages of processing even prior to cooking. Raw meats, uncooked grains, vegetables and even fruits, require preparation and often heating to enhance their flavour, rendering their components edible and digestible. The processing and cooking stages also reduce harmful organisms and parasites of public health significance.

Many technologies may be used to preserve foodstuffs. Pasteurisation deactivates spoilage enzymes and reduces bacteria through heating to temperatures of 80 to 90°C. Heat sterilisation uses atmospheric steam at 100°C for high-acid foods, and pressurised steam at around 120°C for low-acid foods. Irradiation destroys micro-organisms by destruction of their DNA through ionising radiation. These modern techniques use increasing amounts of energy, depending on the level of preservation required.

However, simpler methods are available for the creation of unfavourable conditions for microbes such as:

- Dehydration or reduction of available moistures to between 8 and 12 per cent;
- Freezing, which is in effect moisture immobilisation which reduces microbe and enzyme mobility;
- Salting or syruping which results in osmotic removal of moisture required for life processes;
- Fermentation which inhibits the growth and numbers of microbes by means of alcohol, acid, and sugar or salt concentrations;
- Pickling or smoking with combinations of salt, acid and smoke pasteurisation to reduce microbes; and
- Combinations of two or more of the above methods, in conjunction with irradiation to inhibit microbes.

As many of the above processes require heat they often entail the use of already scarce traditional energy sources, thus placing an increased burden on energy-poor rural households. Thus a number of food preservation technologies work without commercial energy inputs.

Energy in Rural and Agro Industrial Sector

In general, rural industries can be broadly classified into agro and non-agro-based industries. The former would consist of facilities such as those for tobacco curing and rice milling, while the latter would include charcoal and brick manufacturing facilities, potteries, bakeries, blacksmithy, and carpentry and village workshops. Shops and establishments that do not fall under either of these categories could be classified as the rural services sector.

The use of energy for income-earning activities is often substantial and is not easily distinguishable from pure household consumption, or it may simply not be measured. Examples include beer brewing, boiling sugar from cane, pottery, tobacco and copra drying, blacksmithy and baking. Often these goods are produced for own consumption as well as for sale. The scale of error that could arise if these energy uses are not taken into account is illustrated by the results of a survey in rural Bangladesh. For landless families, annual per capita consumption for all kinds of cooking and food preparation was found to be 6.9 GJ/year, of which 6.6 GJ was for domestic cooking. The small remainder was for parboiling rice and making 'ghur', or sugar syrup. For the largest farmers, the equivalent figures were 16.3 and 8.3 GJ/year. The latter used more than twice as much energy in total, but little more than the landless poor for domestic cooking.

The energy needs of rural industries comprise lighting, process heat, motive or shaft power and, increasingly, electricity for computers and communication. Lighting requirements are invariably met by electricity, either from the grid or from diesel or petrol generators, or by kerosene in unelectrified villages. Motive power requirements are met by electricity (where it is available) or by diesel engines or human labour using mechanical equipment, where electricity is not available. The principal supply sources for process heat for blacksmithy, brick making and charcoal manufacture are woodfuels and other biomass. In agro-based facilities, the use of biomass is widespread.

With increases in rural electrification and greater availability of other commercial energy sources in the rural areas, a steady transition from traditional to commercial sources of energy is found. However, traditional fuels remain the major source of process heat in all developing countries, with their overall share of the sectoral energy consumption ranging between 60 and 80 per cent in Asia.

Strategy for Sustainable Rural Energy Supply

Meeting the energy needs of the poor and dispersed rural communities with targeted actions is a major challenge. The rural energy problems are intertwined with many aspects of rural life, which is characterised by socio-cultural, traditional, agrarian, and non-commercial aspects, that are often distinctly different from those in urban areas, and thus appear to be particularly daunting. The challenge to improving the rural energy situation including modernisation of rural energy services lies in:

(i) securing a policy priority at the national level for rural energy modernisation and removing the urban bias;
(ii) improving access by the rural population to commercial fuels, especially liquid and gaseous fuels;
(iii) enlarging energy choices for basic needs, such as cooking and lighting;
(iv) promoting diverse options for rural electrification, including decentralised systems;
(v) developing rural energy markets;
(vi) financing decentralised energy systems;
(vii) improving service and maintenance skills; and
(viii) integrating energy advances with other aspects of rural development.

Options available to India to address rural energy problems cover:

(i) improving the quality and quantity of information on the availability and use of energy from all sources in rural areas for use by policy makers;
(ii) integrating energy policies into overall rural development strategies with emphasis on income generation;
(iii) initiating and/or strengthening rural development policies, including the elimination of bias towards urban development;
(iv) establishing a high priority for the provision of electricity in a cost-effective manner to unserved rural populations;
(v) promoting sustained programmes of investment in decentralised energy schemes based on renewable energy for rural areas not connected to grids wherever such schemes are cost-effective;

(vi) adopting rational pricing of energy to encourage conservation and efficient use;
(vii) removing, in phases, subsidies and/or poorly designed price controls;
(viii) introducing innovative financing arrangements in rural areas, especially for low-income inhabitants, such as micro-financing and cooperative arrangements with credit and licensing agreements to encourage the private sector to provide certain energy services in rural areas;
(ix) raising awareness and educating consumers about the advantages of renewable energy technologies by utilising and/or strengthening existing extension programmes;
(x) disseminating information; and
(xi) enhancing skill and training.

The Challenge—Make Energy Affordable to Poor

In many cases the rural poor are willing and able to pay for energy services if appropriate financing options are offered. The economics of providing basic electricity to rural household should be evaluated according to the cost of supplying comparable energy services through less efficient carrier. The challenge is to find ways to make modern energy carriers affordable to satisfy the basic needs of all rural residents — which may, at least initially, require subsidy. The key is to introduce market efficiency if possible. To use the smallest subsidy, needed to achieve social objectives. If a subsidy is required, it might be provided as an integral part of a new socio-contract, whereby energy provider serve rural energy needs while simultaneously, highly competitive conditions are created in the energy sector (a key element of energy reforms).

Advanced Energy Technologies

Fossil energy: Sustainability goals indicate the importance of evolving fossil energy technologies towards the long-term goal of near zero-air pollutant and greenhouse gas emissions without complicated end-of-pipe control technologies. Near-term technologies and strategies should support this long-term goal.

The technological revolution under way in power generation, where advanced systems are replacing steam turbine technologies, does support this long-term goal. Natural-gas-fired combined cycles offering low costs, high efficiency, and low environmental impacts are being chosen wherever natural gas is readily available—in some countries even displacing large new hydropower projects. Cogeneration is more cost-effective and can play a much larger role in the energy economy—if based on gas turbines and combined cycles rather than on steam turbines.

Reciprocating engines and emerging micro turbine and fuel call technologies are also strong candidates for cogeneration at smaller scales, including commercial and apartment buildings. Coal gasification by partial oxidation with oxygen to produce syngas (mainly carbon monoxide and hydrogen) makes it possible to provide electricity through integrated gasifier combined cycles (IGCC) plants with air pollutant emissions nearly as low as for natural gas combined cycles. Today power from IGCC cogeneration plants is often competitive with power from coal-steam-electric plants in either cogeneration or power-only configurations.

Although synthetic liquid fuels made in single-product facilities are not competitive, superclean syngas-derived synthetic fuels (such as synthetic middle distillates and dimethyl ether) produced in polygeneration facilities that make several products simultaneously may soon be. Syngas can be produced from natural gas by steam reforming or other means or from coal by gasification using oxygen, as noted. Expanding markets for clean synthetic fuels are likely to result from toughening air pollution regulations. Synthetic fuels produced through polygeneration will be based on natural gas if it is readily available. Synthetic middle distillates so produced are likely to be competitive where low-cost natural gas is available (as at remote developing country sites); the technology might facilitate exploitation of relatively small remote natural gas fields.

In natural-gas-poor, coal-rich regions, polygeneration based on coal gasification is promising. Such systems might include production of extra syngas for distribution by pipelines to small-scale cogeneration systems in factories and buildings—making possible clean and efficient use of coal at small as well as large scales. Rapidly growing polygeneration activity is already under way in several countries based on the gasification of low-quality petroleum feedstocks—activity that is helping to pave the way for coal-based systems.

Barriers to widespread deployment of advanced cogeneration and polygeneration systems are mainly institutional. Most systems will produce far more electricity than can be consumed on-site, so achieving favourable economics depends on being able to sell co-product electricity at competitive prices into electric grids. Utility policies have often made doing so difficult, but under the competitive market conditions towards which electric systems are evolving in many regions, cogeneration and polygeneration systems will often fare well.

The near-term pursuit of a syngas-based strategy could pave the way for widespread use of hydrogen (H_2) as an energy carrier, because for decades the cheapest way to make H_2 will be from fossil-fuel-derived syngas. The successful development of fuel cells would facilitate the introduction of H_2 for energy. Fuel cells are getting intense attention, especially for transportation, because they offer high efficiency and near-zero air pollutant emissions. Automakers are racing to develop fuel cell cars, with market entry targeted for 2000-10. The fuel cell car will compete for the role of 'car of the future' with the hybrid internal combustion engine/battery powered car already being introduced into the market.

Syngas-based power and H_2 production strategies also facilitate separation and storage of CO_2 from fossil energy systems, making it possible to obtain useful energy with near-zero emissions of greenhouse gases without large increases in energy costs. Recent research suggests that the global capacity for secure disposal of CO_2 in geological reservoirs might be adequate to dispose of CO_2 from fossil fuel used for hundreds of years, although more research is needed to be sure about this.

Other advanced technologies (ultrasupercritical steam plants, pressurised fluidised-bed combustion, coal IGCC based on partial oxidation in air for power generation, direct coal liquefaction for synthetic fuels production) offer benefits relative to conventional technologies. But unlike syngas-based technologies, pursuing such options in the near term would not offer clear paths to the long-term goal of near-zero emissions without significant increases in costs for energy services.

Nuclear energy: Worldwide, nuclear energy accounts for 6 per cent of energy and 18 per cent of electricity. Although nuclear energy dominates electricity generation in some countries, its initial promise has not been widely realised. Most analysts project that nuclear energy's contribution to global energy will not grow—and might decline during the initial decades of the 21st Century. Nuclear power is more costly than originally projected, competition from alternative technologies is increasing, and there has been a loss of public confidence because of concerns related to safety, radioactive waste management, and potential nuclear weapons proliferation.

But because nuclear power can provide energy without emitting conventional air pollutants and greenhouse gasses, it is worth exploring if advanced technologies could offer simultaneously lower costs, boost public confidence in the safety of nuclear reactors, assure that peaceful nuclear programmes are not used for military purposes, and demonstrate effective nuclear waste management practices. Unlike Chernobyle-type reactors, the light water reactors (LWRs) that dominate nuclear power globally have a good safety record—although this record has been achieved at considerable cost to minimise the risk of accidents.

The potential linkage between peaceful and military uses of nuclear energy was recognised at the dawn of the nuclear age. Efforts to create a non-proliferation regime through the Nuclear Non-Proliferation Treaty and a series of regional treaties, controls on commerce in nuclear materials and goods and services that might be used to further military ambitions, and safeguards applied to nuclear materials in peaceful nuclear applications have been largely successful in separating peaceful and military uses. If there is to be an energy future in which nuclear power eventually contributes much more than at present, stronger institutional measures will be needed to maintain this separation. These measures should be complemented by technological advances aimed at limiting opportunities to acquire nuclear weapons under the guise of peaceful nuclear energy applications and to steal weapons-usable nuclear materials.

Reactor development activity for the near term has involved both evolutionary LWRs and new concepts. Reactor vendors now offer several revolutionary LWRs with improved safety features and standardised designs, for which there can be a high degree of confidence that performance and cost targets will be met. Another evolutionary activity involves modifying LWRs to make them more proliferation resistant through a denatured uranium or thorium fuel cycle. One concept being revisited, the pebble bed modular reactor, offers the potential for a high degree of inherent safety without the need for complicated, capital-intensive safety controls. A pebble bed modular reactor could also be operated on a proliferation resistant denatured uranium or thorium fuel cycle.

Access to low-cost uranium supplies could constrain nuclear power development based on LWRs. The plutonium breeder reactor, which requires reprocessing spent fuel to recover plutonium for recycling in fresh fuel, was once thought to be a viable option for addressing this challenge. But electricity costs for breeders would probably be higher than for LWRs, at least until late in the 21st Century, and preventing proliferation is much more challenging with reprocessing and plutonium recycling than with LWRs operated on once-through fuel cycles.

Other long-term options for addressing the nuclear resource constraint are alternative breeder concepts—including particle-accelerator-driven reactors, uranium from seawater, and thermonuclear fusion. The prospective costs, safety, and proliferation resistance features of such alternative breeder concepts are uncertain, and the concepts would take decades to develop. Recent research suggests it might be feasible, at relatively low cost, to extract uranium from seawater. Where is concentration is low but total quantities are vast. If the technology could be deployed at globally significant scales, it might be feasible to avoid making major commitments to nuclear fuel reprocessing and plutonium recycling. Fusion could provide an almost inexhaustible energy supply, but it will probably not be commercially available before 2050.

Radioactive waste by-products of nuclear energy must be isolated so that they can never return to the human environment in concentrations that could cause significant harm. Although the safety of long-term waste disposal has not been proven, the technical community is confident that this objective can be realised—largely because of the small volumes of wastes involved. But in most countries there is no social consensus on the goals and standards for radioactive waste disposal and on strategies (both interim and long-term) for implementing them. The issues involved are only partly technical. The current social stalemate on waste disposal not only clouds prospects for nuclear expansion; it also has made spent fuel reprocessing a *de facto* interim nuclear waste management strategy in some countries. This has happened even though fuel reprocessing does not offer economic gains and does not solve the waste disposal problem—it merely buys time and is creating large inventories of plutonium that must be disposed of with low proliferation risk.

11

Networking of NGOs in Renewable Energy Use

Energy issue is directly linked to social development in developing countries. The reason for an emphasis on social development is that in most of the developing countries, people in the rural areas do not have even the minimum of the facilities relating to shelter sanitation and energy. The Johannesburg Environment Summit Conference focused on the *Indivisibility of Human Dignity* and resolved through decisions on targets, timetables and partnerships to speedily increase access to basic requirement such as clean water, sanitation, adequate shelter, energy, health care, food security and the protection of bio-diversity. At the same time, the Summit also resolved to work together to assist one another to have access to financial resources, benefit from the opening of markets, ensure capacity building, use modern technologies to bring about development and make sure that there is effective technology transfer for human resource development. The Summit Declaration on Sustainable Development also indicates the commitment of world leaders to ensure that women's empowerment and emancipation, and gender equality are integrated in all activities encompassed within Agenda-21, the Millennium Development Goals and the Johannesburg Plan of Implementation.

The Plan of Implementation clearly brings out the need to integrate energy considerations including energy efficiency, affordability and accessibility, into socio-economic programmes, especially into policies of major energy consuming sectors. It also focused on developing and disseminating alternative energy technologies with the aim of giving greater share of the energy mix to renewable energies, improving energy efficiency and greater reliance on advanced energy technologies, including cleaner fossil fuel technologies. The Plan of Implementation

---

Paper presented at the Workshop on Networking of NGOs in Renewable Energy, organised by the All India Women's Conference on December 7, 2002 at New Delhi.

also emphasised on developing and utilising indigenous energy sources and infrastructure for various local uses and promote rural community participation, including local Agenda-21 Groups, with the support of the international community, in developing and utilising renewable energy technologies to meet their daily energy needs to find simple and local solutions.

Networking between Centres of Excellence in energy sector leading to sustainable development was also emphasised.

Without access to energy services, people must spend a great deal of time and physical energy on basic subsistence activities rather than on earning money. In addition, lack of energy correlates closely with many indicators of poverty, such as poor education, inadequate health care and hardships imposed on women and children. The social issues linked to energy use include, poverty alleviation, opportunities for women, the demographic transition and urbanisation. Overall limited access of energy services marginalises poor people and seriously limits ability to improve their living conditions. The poor typically spend *a greater portion of their income than the rich do on indispensable energy services,* such as cooking. At the same time, they frequently all compromise on services like lighting and space-eating.

Linkages between Energy and Other Sectors

The Millennium Report of the Secretary General of the United Nations focuses on energy and its linkage with water, biodiversity, health and agriculture. This critical and crucial inter-linkage can help in arriving at the requisite socio-economic condition in developing countries. The examples of critical role of energy in the priority areas are as follows:

- *Energy Supply and Use*
  - — Basic needs of poor unmet
  - — Inadequate pattern of use
  - — High environmental impact
- *Energy and Water*
  - — Potential to provide safe drinking water
  - — Hydropower's environmental impact
  - — Acidification of water bodies
- *Energy and Biodiversity*
  - — Bioenergy production to revive degraded land
  - — Hydropower's impact on species and ecosystems

- *Energy and Health*
  - — Potential for improving health care facilities
  - — Traditional fuel use can harm women's health
  - — Health impacts of outdoor air pollution
- *Energy and Agriculture*
  - — Increased productivity through modern energy sources
  - — Bioenergy as replacement for fossil fuels.

Measures to Improve Energy Gender Linkage

There are certain measures that can help to strengthen the linkage between energy, gender and poverty reduction and include the following:

- Introduction of participatory approaches in which both men and women are involved. Further, include women in energy policy formulation, planning and implementation.
- Designing of women oriented energy technologies which can lead to income generation activities and poverty reduction.
- Encouraging local people especially women to become energy entrepreneurs within their communities for long term sustainability of the technologies.
- Provision of information and training to women in managing energy resources.

The above mentioned sectoral inter-linkages on one hand and overarching linkages with gender and poverty reduction indicate that there are a number of inter-disciplinary and multi-disciplinary capabilities that are desired for attaining optimum results. These capabilities cannot be attained by any single organisation or individual. At the same time the developing countries that have scarcity of energy supply normally also face scarcity of funding also. Any effort of duplication of institutional arrangement or organisations with similar objectives and outlook is basically a wasteful exercises. That obviously leads to the need for developing a networking arrangements which will have inter-sectoral linkages and will avoid any duplication in infrastructure creation.

It is normally taken that the women's organisations have a role in technology dissemination only. That is, however, not true. They are a major stakeholder in technology utilisation on one hand and on providing the feedback on field experience and suggestion for modification in technologies to make them suitable for use. Thus they are an equally important link in technology development. There must be an unshakable linkage between R&D Institutions and womens organisations involved in dissemination of renewable energy systems.

Globally it is suggested that policies need to be based on and linked to field experience. Thus the womens organisations need to have a permanent linkage of providing feedback and working out policy strategies for higher use of renewable sources of energy.

Use of local materials for production of devices and equipments for local use can best be generated through local or regional entrepreneurship, preferably within the same communities. This is again, where women have an important role to play and women entrepreneurs can play a much important role in giving devices that are *human sensitive; rather than technology oriented human insensitive devices.*

The last and important area is of disseminating information and training women on renewable energy technologies for building their capacity. Experience shows that such information is passed on much easily when women trainers are involved, whether within the same community or from outside.

There are certain big organisations that have the capacity to build all kinds of capabilities and have the staff accordingly. But on the other hand there are smaller organisations that try such an effort and find themselves in difficult situation. It is, therefore, best option to develop a network of organisations with different kinds of capabilities, like policy formulation, technology generation, entrepreneurship and training. Different networks with these inherent capabilities can cost effectively provide service to the users. These networks should have the grassroot linkages on one hand and also develop their linkages with the regional and national networks and try to keep upgrading their capabilities by incorporating the outcome of the best practices followed anywhere.

12

Renewables for Rural Energy Needs in South Asia

Introduction

Rural energy supply is a key element in meeting the national energy supply in countries in South Asia. 92 per cent of total energy consumption in Nepal which is from renewable/fuelwood energy sources is an indicator of importance of renewable energy in overall energy supply. Evaluation of the 'scale' and 'nature' of the constraints of renewable energy supply includes local energy source and pattern of energy use in rural areas. That gives an insight into the main intervention and how far they help South Asia to achieve rural energy sufficiency. Energy sector reforms were taken up in some of the countries in South Asia with the hope that people will be able to access energy through market mechanism. It was observed that a vast majority of rural people in South Asia have no role in the market mechanism due to lack of purchasing capacity, and remain market excluded. For them, access to commercial sources of energy, at whatever price, is not 'a dream' but 'a nightmare'. Energy supply to rural population in South Asia will have to be planned in a financially acceptable form to its rural poor population.

Evaluation of Rural Energy Programmes

The World Energy Council in its report "The Challenge of Rural Energy Poverty in Developing Countries" (1999) mentions that following three imperatives are apparent if rural energy poverty is to be significantly reduced:

Paper presented at Sixth World Renewable Energy Congress organised by the World Renewable Energy Network, UK during September 2002 at Brighton, UK.

(i) Rural development in general and rural energy specifically need to be given much higher priority by policy makers.
(ii) Rural energy development must be decentralised to put rural people themselves at the heart of planning and implementation.
(iii) Rural energy development must be integrated with other aspects of rural development."

The Report also states, "The biggest obstacle observed in the rural energy supply sector is that in the past, rural development planning was typically driven by an underlying philosophy of trying to achieve rural economic growth: to move economic activities from the subsistence level and essentially to monetise the rural economy as an engine for improvement in rural living standards. It follow that if rural energy planners and policy makers were preoccupied with economic growth, programme objectives would be set to meet this priority, rather than those defined by the rural poor."

Rural energy policy and planning would thus tend to reflect one of the central biases of development, namely:

(i) priorities and objectives are set by outsiders;
(ii) planning is top-down; and
(iii) attitudes and values embodied in the resultant interventions are imposed on rural people.

This bias would appear to be central to the failure of numerous rural energy development programmes ostensibly designed to meet the needs of the poor on the one hand, and to realise specific development on the other. Promotion of fuel-efficient stoves by outsiders, and their subsequent rejection by rural women, is a case in point, as are the numerous failed woodfuel plantation projects scattered around the world. Tragically, by failing to understand the livelihood strategies, needs and priorities of the poor, these programmes have squandered scarce development resources. This experience will guide future course of action in rural energy supply.

Role of Renewable Energy Sources

Renewable energy technologies, RETs, often offer the most appropriate way of meeting energy demand in rural/remote areas. For example, in places where electricity supply grid is unlikely to reach in the near future, producing electricity locally based on a renewable energy source is often the best approach of meeting local electricity demand.

Considering advantages that RETs offer, and their growing importance because of global climate change concerns, all the countries in South Asia have taken concrete steps to promote commercialisation/ diffusion of these technologies. In certain countries specialised entities have been created specifically for this purpose, for example:

(i) a separate ministry in India.
(ii) funding bodies in India and Bangladesh.
(iii) Alternative Energy Promotion Centre in Nepal.
(iv) incentive/subsidy schemes, educational programmes and R&D units in different countries.

It will be interesting to observe the status of renewable energy use in Nepal, Bangladesh and Sri Lanka — the countries truly in transition towards renewable use. These three countries are chosen because they have a systematic approach to development of renewables like solar, wind, biomass (other than traditional use). India has a well developed programme and therefore not described here. Strategically it is better to focus on countries in transition with regard to renewable energy application.

Nepal

In May 1985 a Sherpa started operating a Solar Home System (SHS) in the Solu-Khumbu Region. The area includes the main road to the Sagarmatha National Park which is used by many tourists and expeditions in crossing the ridges towards the Everest base camp.

The commercial production and marketing of flat plate solar collectors for domestic water heating is becoming a mature industry with more than forty manufacturers in the field. The installed capacity of Domestic Solar Water Heaters, DSWHs, is estimated to be around 30,000 litres. Government organisations and many NGOs are also active in the dissemination of the technology use.

A central photovoltaic system was installed (220 V stand-alone AC grid) in Kodari/Tatopani near the Tibetan border in 1987. At that time it was handed over directly to the Nepal Electricity Authority.

The application of solar photovoltaic power has been largely to specific sectors such as communication repeater stations, water pumping stations (for demonstration) and village power systems. An estimated 1 MWp of PV power is currently installed by various public and private sectors in the country in 1998. There is considerable interest in this technology with three manufacturers currently involved in the assembly of imported PV components. The Pulimarang Pilot Project (1994) which has electrified a rural mountainous village (50 SHS) in west Nepal has paved the way for more such projects in the future.

Development of micro hydropower is quite favourable because of the unique hills, scattered settlements and more than 6,000 rivers that criss-cross the country. Nepal has a total of 933 units of micro-hydro turbines with an installed capacity of 9 MW, the majority of which are used for agro-processing. The total number of turbine units installed for producing electricity is 311, with a capacity of around 2.7 MW. There are 11 water turbine manufacturers in the kingdom. Depending on the specific site, the cost per kW varies between US$ 1200 to 1600.

Bangladesh

Commercialisation of renewable energy technologies is a new phenomenon in Bangladesh. Biomass based renewable energy technologies have been practiced in Bangladesh for several decades now. These include improved cooking stoves and biogas plants. Although attempts in the extension have been quite persisting, commercialisation did not start. The same is true for solar thermal systems of which research and demonstration plants, locally fabricated, have been quite diverse. But, none has been carried forward towards commercialisation.

Rural Electrification Board (REB) and Grameen Shakti (GS) have electrified homes and shops in several villages. The total capacity of systems installed by REB and GS are 62 kWp and 22 kWp respectively. The total installed capacity in the country as a whole has been reported to be about 150 kWp in 1999. A pilot project of total capacity 62 kWp has been installed with French and European communities' assistance to provide electricity services to about 1300 consumers. GS has initiated a solar home system dissemination programme with 20 demonstration units, using 17 Wp and 25 Wp modules. There are at least three local companies involved in the supply and marketing of PV products in Bangladesh.

Another technology which made some headway towards

commercialisation in the recent years, almost totally by the initiative of the local small entrepreneurs is, rice husk briquetting.

Recently the production and use of briquettes have increased sharply in certain areas of the country. This is being used in the commercial and industrial sectors as well as in household creating a new use for the husk.

Bangladesh Institute of Technology (BIT), Khulna, conducted a survey and observed that all the machines are of screw type and are locally fabricated. Rapid screw wear has been the main problem. In some places technicians are addressing this problem. Assuming an average production capacity of 90 kg per hour (the lowest type is 75 kg/hr and the highest 150 kg/hr), there is a total production capacity exceeding 81 tones per hour. Though this is not a big figure compared to the wood-fuel consumption in the country, briquetting has exploited potentials in providing a convenient and energy-efficient fuel for the country, if basic technical and commercialisation problems are solved.

Sri Lanka

Sri Lanka embarked on the promotion of photovoltaic for rural domestic use in 1980 and since then many private sector companies have operated with a varying degree of success. A rural electrification project was initiated in 1987 to provide PV systems to 1000 homes in Pansiyagama village; the systems were installed in 1991. An infrastructure project to install 74 large scale systems for rural hospitals consisting of lighting, water pumping, refrigeration, etc. was also funded in 1993. There were about 4,500 solar hot water systems in use in 1996.

The above country reports indicate awareness and acceptability of renewable energy based systems in South Asia.

The recognition of the importance of technology transfer is not a recent development. The Asian and Pacific Centre for Transfer of Technology (APCTT) was established in 1977 in India to promote transfer of technology to and from small- and medium-scale enterprises (SMEs) in Asia and the Pacific. The objective of APCTT is to strengthen the technology transfer capabilities in the region and to facilitate import/export of environmentally sound technologies to/from the member countries. APCTT implements development projects funded by international donors aimed at strengthening the environment for technology transfer among SMEs in Asia and the Pacific. Technologies handled by APCTT include RETs, *e.g.* solar thermal, PV, biomass-based co-generation, biogas etc.

Renewable Energy: A Big Business Opportunity

Renewable energy supply in developing countries in South Asia has been necessitated by unmet demand for energy in the need for diversification of energy supplies, as well as by the increasing deleterious environmental impact of fossil fuel use. In view of the limited resource endowment and the potential of renewables to meet the energy needs in South Asia, particularly of their vast rural population, there is need for clear policy and strategies to be evolved in developing countries to facilitate and enable optimal commercial exploitation of their renewable energy sources rather than exploitation of commercial sources of energy. The development of renewable energy in these countries give benefit and get accelerated by strategic international cooperation and assistance, including at regional and inter-country levels.

Renewable energy has already become a big business opportunity for the enterprising professionals. Low investments in manufacturing facility is a major attraction. And therefore, those enterprising persons who understand the need and future prospects of renewables have started ventures in the renewables in their countries. India has created a reasonable manufacturing base for all renewable energy devices. Bangladesh has specialised in small scale rice husk based briquetting machines. India has created a reasonable manufacturing base for all renewable energy devices.

Lack of financial resources is the major obstacle in wider manufacturing base in each country. Nepal and Bangladesh have started assembly of solar water heating systems. Sri Lanka has enterprising businessmen interested in the manufacture, but with their own resources.

A number of studies have been recently conducted to ensure smooth passage in the age of renewables.

Way Forward

Following inferences will indicate the way to move forward in use of renewables:

(i) Over two billion people, which represents one-third of the global population, live without access to commercial forms of energy. About one billion of this population is in the

South Asia region. This energy poverty is a major concern for social and economic development of these communities.

(ii) A significant number of commercially attractive applications for renewable energy systems have been identified (This is in addition to traditional use of biomass as fuelwood.) Some are matured technologies, many can best be classed as special situations, where local conditions such as remoteness from conventional energy supplies, severe environmental conditions, small size, or special incentives have created conditions favourable for application of renewable energy sources.

(iii) Some applications observed to be viable on lifecycle measures, face the major hurdle of high capital investment. This turns out to be a disadvantage when initial cost is the major criteria used for energy system selection.

(iv) Overall cost of energy supply from renewable energy systems has reduced over the last decade, due to higher production volumes and incremental improvement in efficiency, materials reliability and application.

(v) Severe environmental, economic and technological challenges are surfacing to tackle carbon dioxide and other greenhouse gas emissions. Renewable energies have an important place for their environmental benignness and can play important role in retarding the climate change effect due to fossil energy use.

(vi) Modular nature of many of the renewable energy technologies is beneficial for mass production of standard sizes/models and, contribute to a relatively short planning and implementation cycle.

(vii) Data on renewable resources have been collected and experience has been gained over the last 20 years. A number of pioneering R&D and demonstration projects have been developed in the region in collaboration with developed countries and international agencies. The commercially viable options so developed have evolved in the local conditions and are suited to the needs of the region.

(viii) Research and Development on emerging renewable technologies needs to be expanded so that these options can be rapidly moved to maturity and commercialisation. Private industry participation from Conception State can reduce time lag in commercialisation of technologies.

(ix) Electricity production from renewables, both for grid connectivity as well as for stand-alone has become a reality. Major initiatives in some of the countries in the region are sufficient to build confidence in this approach and multiply their numbers.

(x) Lack of infrastructure and institutions for promotion of renewable energy applications needs to be overcome by strengthening/upgrading existing institutions and establishing new institutions to promote public and private funding in this sector.

(xi) Sharing of 'best practices' in the region is a cost-effective route. A number of successful packages can be developed through funding from international agencies like the UNDP.

(xii) Technology transfer, obligatory for developed countries under the Kyoto Protocol, can be effectively bartered by the developing countries for continuous upgradation of their facilities.

(xiii) Short term planning and economic decision making is a major obstacle to renewables, which needs to be replaced by long term planning based on resource data and technology proving.

(xiv) A need is felt for effective coordination of government's sectoral plans dealing with energy, agriculture, forestry and rural development. Major initiatives are beginning in non-governmental organisations and the international funding organisations to provide leadership for increased renewable energy use.

(xv) An Apex Organisation for coordination amongst industries in the region can promote flexibility in operation, including technology transfer.

(xvi) Present priority for renewables by various governments, as reflected in their policy announcements, needs much more effective cooperation at the inter-governmental level, by involving the international agencies, funding agencies, industries and NGOs.

Conclusion

Regional cooperation in countries of South Asia is taking shape with the involvement of governments, semi-government agencies, private industries and NGOs. Rural energy supply in South Asia is not possible by fossil fuels or centralised grid connected electricity systems only. Renewable energy applications based on local grid connected and stand above systems can meet the rural energy needs in South Asia. Rapid commercialisation in renewable energy systems is the best strategy to move away from the subsidy-pushed devices to self-supporting systems. This course is not easy, but desirable. Attaining this goal will call for reduction in rural poverty. The 'rural energy poverty' can be overcome through renewable energy systems made available to the 'rural poor' at an affordable price.

REFERENCES

Annual Report 1998-99, Ministry of Non Conventional Energy Sources, Government of India.

Chaturvedi, Pradeep, (1998). *Regional Workshop on Rural Energy Supply*, July 10-11,1998, Colombo. September 21-25, 1998.

Chaturvedi, Pradeep, (1999). *Energy Management — Challenges for the Next Millennium*, Concept Publishers.

Chaturvedi, Pradeep, (1999). *Environment Friendly Energy Supply through Biomass in the Next Millennium*, World Renewable Energy Congress' 99, June 8-12, 1999, Kuala Lumpur.

Country Paper — Bangladesh, Workshop on Commercialisation of Renewable Energy Technologies for Sustainable Development, ESCAP, January 11-12, 1999, Bangkok.

Country Paper— Nepal, Workshop on Commercialisation of Renewable Energy Technologies for Sustainable Development, ESCAP, January 11-12, 1999, Bangkok.

Economic Survey of India 1999, Government of India.

National Paper: India's Energy Scenario 1998 — Presented at the 17th Congress of WEC at Houston, September 13-18,1998.

Report of the Regional Workshop on Commercialisation of Renewable Energy Technologies for Sustainable Development, ESCAP, January 11-12, 1999, Bangkok.

13

Regional Cooperation for Sustainable Energy Supply

Introduction

The Nepal Engineers' Association have decided on 'Engineering Profession in National Building: Construction and Vision', a very appropriate theme for 7th National Convention. The role of engineers in nation building has been often overlooked. They are marginalised to carry out the orders and provide professional inputs to support the framework of the government policies. They are often not a part of the decision making body. Even when the globe has moved from the 'industrial age' to 'information era', based on the knowledge economy, mainly due to the role played by the engineers alone whether in terms of development of theory for knowledge based systems or for dissemination of such systems, they do not find their rightful place. By focusing on the theme it will be possible to discuss the status.

The paper is not intended to deliberate on potential of hydro-power and the specific advantages of developing the same and as such issues of significance leading to policy, implementation, regional cooperation are only dealt in this paper.

Globalisation and South Asia

The concept of globalisation has caught on in the South Asia region and every country is deeply concerned about how best can the advantage of globalisation be taken to ensure higher exports and foreign direct investments. Irrespective of the needs and the challenges before the countries in the South Asia region, the foreign investment is not

---

Paper presented at the 9th Annual Convention of the Nepal Engineers' Association, during April 9-11, 2001 at Kathmandu.

available at the desired pace. Every foreign investor expects mature domestic capital markets before they invest in energy projects. It is necessary to look at the countries in South Asia as one region and discuss about the development of capital markets in the South Asia region. The capacity of the South Asia region to guarantee the foreign investors the return on their investments will be the major factor in speedy development.

Nepal is strategically situated in South Asia and with its core competence in service sector — mainly the tourism — and rich endowments of natural renewable resources, it has the flexibility to develop energy sector at a fast pace to fulfil the energy needs; and to export surplus energy to earn foreign exchange or access services.

Linkage between energy and major global issues relates to decisions that occur at a number of levels. At one level, it is important to consider how best to resolve the issue of many basic energy needs. At another level most developing countries face the challenge of sustaining rapid increases in energy supplies and energy services needed for their development. The issue that needs to be addressed in a comprehensive manner with due consideration to environmental and health aspects. At yet another level, it is equally important to address the issue of energy security and market stability, as well as the issue of many long term global demands for energy. Energy security issues have to be viewed with respect of supply side, as well as demand side concerns.

Challenges to Improving Accessibility of Energy

Access to energy is crucial to the achievement of economic and social development, and alleviation of poverty, as well as to finding an appropriate response to environmental concerns. Improving accessibility of energy implies finding ways and means by which energy services can be delivered reliably, affordably and in an environmentally sound and socially acceptable manner, particularly in meeting the energy requirements of developing countries, including their rural areas.

Furthermore, accessibility of energy is conditioned by security of supply and demand as well as by market stability, all of which need to be addressed in terms of their regional and international dimensions, which include cross-border energy transport infrastructure, storage, and shipment by tankers.

The South Asia has a atypical characteristic of heavy dependence on biomass fuels, which is expected to continue well up to the year 2020. Countries report to be meeting most of their energy needs through biomass — Nepal having reported 92 per cent of its supply from biomass in 1996. However, the modern forms of energy are desirable for rapid growth; and as and when that switchover is considered the situation of the oil and gas sector have to be analytically observed.

Although concern about security of energy supplies is not as intense today as it was in the 1970s and the early 1980s, the availability of future supplies of oil and natural gas continue to preoccupy governments in countries heavily dependent on imports. The full implications to energy security of these developments are unpredictable at this stage. The challenge to enhancing energy security lies in diversifying the energy mix available to consumers.

Thus, challenges to improving accessibility of energy include the:

(a) lack of requisite capacity at national and local levels to address the enormous magnitude and scale of poverty in a number of developing countries, particularly in rural areas without access to modern energy services to meet their basic and development needs;
(b) frequent supply and demand imbalances, and consequent market instability resulting from inadequate information exchange and international dialogue between producers and consumers of energy on supply and demand issues;
(c) technical, economic, social and political challenges to expanding the supply of electricity to the large population in rural areas that are still without access to electricity
(d) barriers to cross-border interconnection of electric grids, cross-border gas pipeline grids, regional underground storage, trade and transit agreements for transportation of energy products, shipment by tankers and market stability, all of which also have regional and international dimensions.

Options and Strategies to Improve Accessibility of Energy

At the national level, focus on satisfying the energy needs of the poor with modern fuels and technologies has the potential to improve standards of living and health, and create new jobs and business opportunities. Addressing energy needs of the poor in developing countries will require not only major macro-economic changes, but also targeted social policy actions that will make commercial energy more widely available. For the rural populations in many parts of the developing world that continue to be heavily dependent on fuelwood and other biomass resources, options that could be considered are:

(a) improving access to sustainable biomass and fuelwood supplies and commercialising biomass operations, including the use of agricultural residues;
(b) progressively shifting to other liquid and gaseous fuels;
(c) developing locally available energy resources for greater energy security through diversification; and
(d) supporting electricity services based on grid extension and/or decentralised energy technologies.

Option (a) above implies that national and local level actions are needed to promote:

(i) fuelwood conservation measures,
(ii) more effective use of agricultural residues,
(iii) cultivation of fast-growing species of plants, and
(iv) use of modern technologies for the conversion and efficient utilisation of biomass, including the use of improved stoves, biogas and solar cookers in households for cooking.

Option (b) implies less dependence on biomass as it seeks to improve access to liquid and gaseous fuels by households particularly rural households. This will have a major effect in reducing indoor air pollution and improve the health of women and children. Such a shift will require economic liberalisation policies that would make it possible for wholesale and retail marketing of these fuels and that would also promote the sale of corresponding appliances by entrepreneurs.

Option (c) implies policy support for the development and utilisation of local energy resources, particularly renewable resources that can improve accessibility of energy. This action can also reduce burdensome dependence on imports thereby contributing to greater energy security.

Option (d) emphasises the electrification of rural areas through grid extension or local grids with decentralised alternatives to improve accessibility of electricity services.

Policy options, an adequate and reliable supply of electric power lie in creating an enabling environment for the private sector or energy cooperatives to engage in the generation and distribution of electricity on a commercial basis including decentralised electrification solutions based on solar, wind, biomass, small hydro and their combinations with other alternatives.

Rural Energy Supply

Rural energy is important for South Asia as that will ensure empowerment of people to overcome poverty.

In case of the rural energy options, there is a growing recognition of the importance of an integrated approach to rural development, and of the linkages between energy, agriculture and environment. Efforts at finding the most appropriate solution of the energy problems of rural areas are hampered by insufficient attention to rural areas and to rural energy needs in particular. Meeting rural energy needs of the poor dispersed rural communities with targeted actions is a major challenge. Modernisation of rural energy services will need attention on the following:

(a) securing a policy priority at the national level for rural energy modernisation and removing the urban bias;
(b) improving access by the rural population to commercial fuels, especially liquid and gaseous fuels;
(c) enlarging energy choices for basic needs, such as cooking and lighting;
(d) promoting diverse options for rural electrification, including decentralised systems;
(e) developing rural energy markets;
(f) financing decentralised energy systems;
(g) improving service and maintenance skills; and
(h) integrating energy advances with other aspects of rural development.

In order to build human resource capacities in this field in rural areas, including the community based organisations and NGOs, targeted actions are needed that cover:

(a) information resources;
(b) participatory approaches;
(c) organisational support;
(d) concrete energy development activities involving appropriate technologies, the provision of programme tools and guidance material for effective management;
(e) training to improve repair and maintenance skills;
(f) entrepreneurial capacity development;
(g) facilitation of local manufacture; and
(h) the establishment credit facilities and support for self-reliant local institutions and networking arrangements.

The Future Course of Sustainable Energy Supply

Energy policies for sustainable development will depend on a range of factors. This include attitude and behaviour, information and technologies, the availability of finance and supporting institutions and in particular policies and policy framework that encourage change in the desired direction. The current path of energy development and the rate of change, are not competitive with key elements of sustainable development. The divergence of alternative features that become apparent in the scenarios after 20 years reflect the long term nature of energy systems.

The most critical issue that sustainable energy strategies and the policies derived from them need to address are how to widen access to reliable and affordable modern energy supplies and how to ease the negative health and environmental impacts of energy use.

Policies and policy frameworks in support of sustainable development need focus on widening access, encouraging energy efficiency, accelerating new and renewable energy diffusion and expanding use of advance clean fossil fuel technology.

The broad strategies for encouraging sustainable energy systems have six major elements:

(a) Making markets work better;
(b) Complementing energy sector restructuring with regulations that encourage sustainable energy;
(c) Mobilising additional investments in sustainable energy;
(d) Encouraging technological innovation:

(e) Supporting technological leadership and capacity building in developing countries;

(f) Encouraging greater cooperation at the international levels;

Actions at the National Level

Actions are needed at the national level to strengthen the capacity of community-based organisations and institutions, including women's groups at the local level, to facilitate participatory approaches to energy for sustainable development. As part of this capacity building effort, assistance needs to be provided in the form of information resources, organisational support, training, planning and management tools, and entrepreneurial skills so as to facilitate access to modern energy services. The international community can play a vital role in catalysing further targeted action at the national levels so that the global scale problem of 2 billion people and the regional scale problem of 700 million still without access to modern energy services could receive much deserved attention. Building local capacity is also vital for the development of local energy resources and deployment of decentralised energy systems, thereby enhancing energy security.

Options and Strategies for Regional Cooperation

At the regional level, the actions that can be considered are the:

(a) development of regional cooperative arrangements for increased regional energy trade in fuels and electricity and investments in energy infrastructure, thereby improving accessibility;

(b) creation of institutional arrangements and regulatory mechanisms for promoting the interconnection of electric grids; and

(c) establishment of dialogue forums between regional producers and consumers for enhancing regional trade.

There is also need for appropriate mechanisms to promote cooperation on issues like energy security, energy treaties, regional storage of fuels, interconnection of electric grids. Arrangements are also needed to facilitate North-South and South-South cooperation.

It is incumbent upon the international community to establish arrangement to assist the least-developed countries in obtaining the energy they need for sustainable development at prices they can afford.

Conclusion

Energy can serve as a powerful tool for sustainable development in the South Asia region. The poverty, inequity, inefficiency, unreliable service, immediate environmental priorities, lack of information and basic skills require changes to be made. These changes will have to be brought around by all the countries in South Asia region as fast as possible to ensure reaching the sustainable energy goal by year 2020.

14

Synergy in Hydro Power Development in Asia

Introduction

River valley development with specific reference to hydroelectric power development gives a new dimension for synergistic development ultimately aimed at hydroelectric power development in Asia. It brings into focus aspect of the multifunctional use of water resources and therefore the need for sharing of financial resource for development of river valley and hydroelectric power projects. This concept may not be applicable to all the countries in Asia, but is a significant for countries in South Asia. To cover whole of Asia it will be appropriate to deal separately with two groups of countries in South Asia and South-East Asia. China with its massive programme in the hydroelectric power development obviously needs a separate treatment. This kind of grouping will help in identifying policies in different sub-regions of Asia. To ensure uniformity of data, reference has been made to the World Energy Council Resource Handbook, 1998. The paper deals with specific developments in 'major' and 'mini and micro' hydroelectric power projects as two separate issues.

Till the early 90s there has been less than 100 GW (about 350 TWh/year) of new hydro capacity under construction at anyone time, equivalent to less than 15 per cent of the capacity in operation. This figure has recently risen, mainly due to China's vast construction programme, which includes the 18,200 MW Three Gorges Project, now in its second phase of construction. Most of new hydroelectricity capacity is under construction in Asia and South America. China has about 50,000 MW under construction; and Brazil, India and Russia each have about 9,000 MW under construction. Hydroelectricity is now considered an important function of many multipurpose water projects.

There is also significant, widespread activity in developing small, mini and micro hydro plants. Atleast 40 countries, particularly in Asia and Europe, have plants under construction and even more have plants planned for construction.

Hydropower development in Asia through mini and micro hydro plants can be realised from the fact that almost all the countries in Asia have such plants under construction. The thrust on local resource utilisation, after the economic setback in South East Asian countries in 1997, has accelerated small hydro development in all Asian countries.

Hydro power is gaining importance in its role in reducing or controlling emissions of so called GHG by estimated 10 per cent through displacement of thermal generation. Environment funds are being allocated in larger quantity to support mini and micro hydro power development in difficult geographic locations. If it were not for environmental reasons, these locations were demarcated as un-economic and impractical.

Potential of some countries exceeds their requirements. This has led to consideration of bi-national or multinational projects. Examples are power exports from Bhutan to India and Laos to Thailand justifying the projects. A number of similar other options have been studied for power transfer from Pakistan to India, Nepal to India and Bangladesh to India.

A new dimension has been added by stiff public opposition to large dams and large hydro power projects for environmental reasons and resettlement of displaced persons from the project site. India, Philippines, Sri Lanka and Thailand have faced such opposition. Several projects have been delayed or halted — Sardar Sarovar Dam and Tehri Dam are two of them.

Recently, however, three things have happened in most ASEAN countries to change the picture. First, there has been a shift in emphasis from the large national enterprise approach to the small and medium enterprise (SME) approach that opened up the private entrepreneurs to independent power producers (IPPs), among which are small power producers (SPPs). Second, the SME approach has led to a market driven, more commercial approach resulting to the expansion of SPPs using a variety of NRE systems based on indigenous resources. Third, the ASEAN region's economic crisis has depressed the demand for electricity in the urban and industrial areas but highlighted the demand growth in the rural off-grid areas which are practically unaffected by the crisis. Thus, rural electrification provides a big opportunity for small power systems based on indigenous renewable energy resources including the small and micro hydro plants.

This is the broad scenario which reflects the potential of development of additional hydro resources, but also highlights the issues other than technical and financial that need to be resolved if total potential is to be harnessed.

Status of Hydro Electric Power Capacities

Hydroelectric power generation is significant for a number of countries for its economic renewable characteristics. It plays an essential role as the widely used renewable source of electricity in over 150 countries in the world. 31 countries in Asia generated and utilised hydro-electricity. A survey in 1997 by the International Journal of Hydropower and Dams found that hydro supplies at least 50 per cent of national electricity production in 63 countries and at least 90 per cent in 23 countries. 10 countries obtain essentially all their commercial electricity from hydro — including Norway several African nations, Bhutan and Paraguay.

About 700 GW of hydro capacity is in operation (in 1996) world-wide, generating 2600 TWh/year (about 19 per cent of the world electricity production). About half of this capacity and generation is in Europe and North America. However, this proportion is declining as Asia and Latin America investing large amounts on creating new hydro capacity.

Small, mini and micro hydro plants (usually defined as plant less than 10 MW, 2 MW and 100 KW respectively) play a key role in many countries, often being the mainstay of rural electrification. An estimated 300 million people in China, for example, depend on small hydro.

Review of Major Hydroelectric Power Projects in Selected Countries

A review of major and minor hydroelectric projects in selected countries — South Asia; South East Asia and China — will highlight the growth pattern and the similarity of approaches. First the major projects are being dealt.

India

India has one of the world's highest levels of theoretical hydroelectric potential, but its technically exploitable capability is on a somewhat different scale. This has been assessed as slightly over 84,000 MW, which at an assumed 40 per cent annual load factor would imply an output in the region of 294 TWh/year. In 1996, Indian hydro plants produced about 69 TWh, and accounted for 25 per cent of electricity generation.

About 10,000 MW hydro capacity is under construction and another 28,000 MW is planned. Plants under construction include Naptha Jhakri (1,500 MW); Sardar Sarovar (1,200 MW); Tehri (1,000 MW); Upper Indiravati (1,000 MW); Indira Sagar (1,000 MW) and Ranjit Sagar (600 MW).

Nepal

Nepal has ideal topography for hydro development, which offers the country its only realistic resource for further economic development. The potential is estimated in the range of 727 TWh/year. The technical potential is assessed at 179 TWh/year and the economic potential is assessed at 43 TWh/year. Output of hydroelectricity in 1996 was only 1.2 TWh from an installed capacity of 254 MW. Currently hydro provides almost all the electricity for Nepal's needs.

Two major hydro projects are under construction: Kali Gandaki (144 MW) and Khimti (60 MW). In view of large investments required Nepal may have to enter into power purchase agreement with India to justify investments from private and foreign funding agencies. However, a firm commitment on power purchase will depend on settlement of the issue of river water sharing.

Myanmar

Myanmar has a total hydro potential of 366 TWh/year, of which the technical potential is about 130 TWh/year. This corresponds to an installed capacity of about 37,000 MW operational at 40 per cent annual load factor. The current installed capacity is 328 MW producing about 1.6 TWh/year in 1996.

Nine major hydroelectric power projects with combined capacity of 1,519 MW will be implemented in the near future. These are: Paunglaung (288 MW), Bilin (240 MW), Kun (84 MW), Yeywa (600

MW), Mon (75 MW), Zawgyi II (12 MW), Zaungtu (20 MW), Nam Mae Sai (8 MW), and Nam Kok (200 MW).

Malaysia

Mini-hydro schemes are based on run-of-river systems ranging from 500 kW to 1000 kW capacity. There are 16 MW of total generating capacity in Peninsular Malaysia; 3 MW in Sarawak; and 5 MW in Sabah. In Peninsular Malaysia the mini-hydro units are owned by utility company, Tenaga Nasional, but the maintenance of these units is contracted out.

Lao PDR

The country has an estimated hydro potential of about 18,000 MW. So far, about 203 MW of the potential has been exploited. At present, there are eighteen (18) IPP hydropower projects, providing additional capacity of 6,326 MW.

The Theun-Hinboun Power Project (THPP) is Lao PDR's major breakthrough in hydropower development. It serves as an example on the future development of hydropower projects in Lao PDR. It is the first project that has showed the participation of the government, investors from the private sector and international financial institutions. The project has installed capacity of 210 MW capable of producing 1, 620 GWh of electricity per year. It is located some 100 km upstream of the Nam Theun and Mekong rivers and about 280 km east of Vientianne.

THPC will export at least 95 per cent of future electricity output to Thailand. The project is expected to support Lao PDR's economic growth significantly, increasing the GDP by almost 8 per cent or to provide Laos with more than US$ 1 billion in royalties, taxes and dividends during the 30-year licensing period.

Thailand

Hydro resources are considerable. At present 23 small hydro projects (from 200 to 6,000 kW capacity) with a total installed capacity of 72.4 MW have been completed. About 71 micro hydropower plants (less than 200 kW) with an aggregate capacity of more than 2 MW have been installed.

Indonesia

The technically exploitable capability for hydro-electric generation is just over 400 TWh/year (corresponding to 74,674 MW installed), of which 10 per cent is considered to be economically exploitable. Hydro output in 1996 was about 12.8 TWh, indicating the possible scope for further development within the feasible potential. Hydro presently meets approximately 14 per cent of Indonesia's electricity requirements.

At end-1996, just over 2,100 MW of hydro-electric generating capacity was under construction and about 1,700 MW was planned for development. The total installed capacity has been reported to be 3,854 MW by the end of 1998.

China

The hydro-electric potential is vast, however, measured: China's gross theoretical capability approaches 6000 TWh/year, while its economically exploitable capability is put at 1260 TWh/year — in both instances, far larger than any other country. Actual hydro output is approaching 190 TWh/year, which provides around 17 per cent of the republic's electricity generation.

Total capacity of plants under construction is about 50 GW, by far the largest amount in the world, at least four times that of any other country. The largest plants under construction are the 1800 MW Xiaolangdi (successful river closure in October 1997), Three Gorges (18200 MW), now entering phase 2 of construction after a successful river closure in November 1997 and Ertan (3,300 MW).

The largest plants planned are Xiluodu (14,400MW), Xiangjiaba (6,000 MW), Longtan (4,200 MW) later 5,400 MW) and Xiaowan (4,200 MW). Two major pumped-storage plants are under construction: Guangzhou II (1,200 MW) and Tianhuangping (1,800 MW) About 10 GW of pumped-storage capacity is planned.

Small Hydroelectric Power Capacity

India

Small hydropower potential in India has been estimated to be about 10,000 MW. Installed capacity of small hydropower in the country by the end of 1998 is estimated at 155 MW involving 216 projects. Another

208 projects aggregating to 230 MW are under various stages of implementation. A UNDP/GEF Technical Assistance project on Optimising Development of Small Hydro Resources in the Himalayan and Sub-Himalayan regions of the country is under implementation. A scheme for portable micro hydel sets of 15 kW unit capacity has been introduced in the north-eastern region. 28 sets have been installed under the scheme, and another 22 sets are under installation.

Nepal

Development of micro hydropower in Nepal is quite favourable because of the unique hills, scattered settlements and a large number of rivers that criss-cross the country. Nepal has a total of 933 units of micro-hydro turbines with an installed capacity of 9 MW in 1996, the majority of which are used for agro-processing. The total number of turbine units installed for producing electricity is 311, with a capacity of around 2.7 MW.

Malaysia

Malaysia has 22 mini-hydro plants of capacity 25-600 kW in 1998.

Indonesia

Small-scale hydro capacity amounted to 63 MW at the end of 1996, with a further 58 MW planned for future installation. As regards the micro hydro the potential has been assessed at 455 MW and the installed capacity up to 1998 has been 22 MW.

Vietnam

The technical potential of small and micro-hydropower (< 10 MW) in Vietnam has been estimated to be about 1800 MW; of this 61.4 MW has been harnessed so far. 3000 family sized hydro turbine generator set of 1 kW or less have been installed in the country by 1998.

China

A relatively high proportion — in the region of 25 per cent — of China's hydro-electric capacity and output is associated with small-scale schemes. In 1996, total hydro generating capacity in the under-10 MW category was about 14,400 MW, with a further 2,500 GW under construction or planned for future development.

Future Potential for Hydro

Most favourable sites for hydropower potential have been exploited and further probable sites are geographically difficult and economically costlier locations — though the unexploited potential is vast mainly in developing countries of Asia, Latin America and Africa, where populations are increasing fast and where there will be the most rapid increase in electricity consumption.

The world's gross theoretical hydro power potential is about 40,000 TWh/year of which about 14,000 TWh/year is technically feasible for development and about 9000 TWh/year is currently economically feasible. The last figure fluctuated drastically, being influenced not only by hydro technology but by the changing competitiveness of other energy/electricity etc.

Only about 18 per cent of the world's technically feasible hydro potential and 28 per cent of the economic potential have been developed so far. Approximately, 6,400 TWh/year, economic potential therefore remains to be exploited (equivalent to about 1,800 GW).

Only 20 per cent of the economic potential has been exploited in Asia which means that 80 per cent of the economic potential still exist in Asia for exploitation which can substantially change the energy scene and economic development in this region. Hydro will thus remain the main source of electricity generation in many of the developing countries in Asia which possess adequate hydro resources.

Factors Affecting Further Hydro Development

The two kinds of factors that affect hydro development — First contains factors conducive to the development of hydro power; and the other mainly the economic, local, environmental and socio-economic factors that work as barriers to hydro development.

Factors Conducive to Development

Following factors are conducive to development:

(i) Its renewable, non-polluting characteristics.

(ii) Resources are widely distributed compared with fossil and nuclear fuels.
(iii) The technology is accessible by all countries and provides energy independence.
(iv) Hydro plants have long service lives, are simple to operate, and have comparatively low maintenance and operational costs.
(v) Hydroelectricity can be generated at a stable, competitive unit cost.
(vi) Hydro plants facilitate optimisation of system operation, providing operational flexibility.
(vii) Hydro plants can form part of multipurpose schemes which provide other important benefits.

Barriers to Development

However, economic, local environmental and socio-economic factors that inhibit hydro development include:

(i) Limited financial resources for the high initial costs of hydro, particularly in the developing countries.
(ii) Perceptions about the environmental compatibility of hydro projects (particularly if large-scale).
(iii) The tendency for shot-term financial planning.
(iv) Increasing pressure on finite freshwater resources (the highest use being for agriculture).

Major Challenges to Hydropower Projects

The hydropower projects have to face two major challenges as follows:

(i) The sites will be located in developing countries, which do not have their own financial resources to fund such projects.
(ii) Environmental lobby is deeply pitted against exploitation of hydro sites either they require large dams or they are in the locations that affect the bio-diversity.

Thus, even though the exploitation of the potential is ultimately advantageous for mitigation of GHG, the up-front hurdles of environmental apprehensions and financing will not allow total exploitation.

Environmental Concerns

Wherever electricity is being generated from fossil fuels, or is an option introduction of a hydro plant offers environmental benefits (notably the avoidance of emissions) through displacement of thermal generation and increased efficiency. Recent lifecycle analysis has concluded that development of half of the world's remaining economically feasible hydro potential could reduce GHG emissions (1990 estimates) by 13 per cent, with an even greater impact on SO_2 emissions.

Those who opposed hydro power on environmental grounds do not take into consideration the GHG emissions mitigation potential. The negative impacts have been sometimes overemphasised by opponents. They assume that the choice is between particular 'hydro project or none at all'; while in reality, the choice is between 'hydro power plant or some other type of power stations, usually thermal'. If all impacts of the lifecycle of other plants were considered, hydro would more often be viable.

An increasing number of countries are establishing a legal framework for environmental assessment and management for water resources, and other projects. Addressing environmental aspects at the planning stage has increasingly become accepted as essential and normal for a successful project, rather than being viewed as an unavoidable, unnecessary legal obstacle. Increasing environmental concern, while generally favouring hydro in relation to other options, can thus also have the effect of slowing down further hydro development.

Financing Aspects

The high initial investment cost associated with hydro project planning and construction make private financing difficult, while government cannot usually hand-over full responsibility for hydro site to a private developer in view of local issues relating to control of water resources, environmental protection and safety.

If the long operational lives of hydro plants, reliability and reduced GHG emissions are taken into account, hydro is often the most attractive power option in the region with sufficient resources. If full account is taken of all the costs of other options, and multipurpose hydro projects are considered (with the costs for non-hydro benefits borne by the direct beneficiaries and not by electricity consumers) – possibly with construction in stages — the investment burden may be significantly reduced. Recent private sector initiatives in hydropower in India, Nepal, Indonesia and Laos is encouraging sign for future development.

The above factors lead us to consider multi-purpose hydro projects and to consider synergy amongst various activities.

Building Synergy

Two possibilities have been observed in two different locations which project potential of better cooperation for energy use in sub-regions. Achieving synergy amongst all countries of Asia region is not practical. Synergy can be effectively developed in countries in South Asia on one hand and the South-East Asia on the other. The reason for this suggestion is that the systems will operate more or less under similar geographical, social and economic situations requiring similar kind of approaches.

Synergy has to be created at the country level and also at the inter-country level. Some of the issues to be considered for this synergy are as follows:

(i) Develop the concept of synergy at the country level between river valley development and hydroelectricity development in developing countries of Asia.

(ii) Identify commonalities of problems, especially with regard to environment and rehabilitation issues.

(iii) Identify approach for amicable agreements on sharing water resources and hydropower amongst different countries.

(iv) Promote Regional Cooperation amongst countries benefiting from water and hydroelectricity resources.

(v) Develop linkages amongst different sectors of national planned development like water resources; energy (including power); agriculture; rural development; environment and forests etc. who are direct stakeholders for water resource use.

Conclusion

Synergy amongst the countries utilising the resources is extremely important to find acceptable solution all crucial issues of development. Hydropower potential in some countries is not being tapped as there is no local consumption and no assured power export agreement. For example, hydropower development in Nepal is possible only if assured market for export, mainly to India, can be established. On the other hand the project development in isolation may lead to large regional imbalances in installed generation capacity without their local consumption and utilisation of power, like in Pakistan. That will also relate to export agreements. Thus, sharing of hydropower potential provides an opportunity for better exploitation of hydro resource but only through cooperation and better synergy in approach.

Developing countries in Asia can best achieve effective levels of hydropower development only through the concept of 'Synergy through Networking'.

PART II

ENVIRONMENT

15

Environmental Issues in Delhi and Action Plan (2000-2002)

I am privileged and honoured to have been elected to the highest office of the Institution of Engineers (India), the Delhi State Centre. I, and my family members, wishes to whole heartedly thank the members for giving me this opportunity. The real challenge lies ahead. I need to uphold the prestige and dignity of the office and also work in the spirit that will fulfil the aspirations of the members — something that can be attained only with the support of all the members and blessings of the Almighty.

Looking into the deterioration in the quality of life of the common man in the city of Delhi, I have chosen to speak on *'Opportunity for the Delhi State Centre to arrest Environmental Degradation in the City of Delhi'*. Some members might wonder as to why do I not speak on Information Technology — which is the 'Haute Couture'. I have always looked at IT as a tool to attain other economic and social goals in a cost-effective manner. The Institution has to integrate IT in its operation to ensure better opportunity for living conditions to the common-man.

City of Delhi — Environmental Issues

I will now like to refer to major problems faced by the city of Delhi.

The Master Plan Delhi, 2001

The situation in the city of Delhi is peculiar, and beyond the realm of urban and city planning. The Master Plan Delhi, 2001 has become outdated. The Master Plan projections went awry as the exponential

---

Address at the installation as Chairman, The Institution of Engineers (India), Delhi State Centre on Novemebr 13, 2002 at New Delhi.

growth of man and machine in the city was far beyond expectation; and the spread of human settlements in rural Delhi grew at a much faster pace than projected. The urban slums and urban villages have mushroomed and overstressed the civic system. Not to blame the urban planners, who could plan only for expected growth rates, the situation needs solutions outside the normal planning tools.

With rapid population increases over the past 30 years, since the First Environment Summit in 1972, the environmental quality of both rural and urban areas has continued to deteriorate.

Urban population faces the problems due to rapid urbanisation, such as land use; urban air and water pollution and associated diseases; non-availability of safe drinking water and the overexploitation of groundwater causing urban groundwater subsidence; increasing traffic congestion, noise pollution and significant increase in solid municipal and industrial wastes. The urban poor are the most affected by urban environmental degradation. They tend to live and work in the most polluted and environmentally sensitive places and they do not have the means to protect themselves against pollution. The urban slums exist in pathetic environmental conditions that affect the health of the poor and further impoverishes them.

The *major environmental problem facing the rural population* are those associated with '*poverty*' and '*underdevelopment*', such as access to drinking water, sanitation, education and health care, poor rural infrastructure for agriculture and related developments, and poor nutritional intake. Increased use of fertilizers and pesticides is contaminating the land and water resources.

Air Quality

Industrialisation, poor quality of fuel, and inefficient and insufficient public transport systems are the factors behind the increasing air pollution. Inefficient and unsustainable energy consumption patterns in industry, households and transport have also contributed to air pollution. According to some estimates, automobile and industrial emissions account for two-thirds of the air pollution. The situation in the city of Delhi is alarming and, barring for marginal improvements in the quality of air, desirable improvements have not taken place.

Particulate Matter

The city of Delhi exceeds World Health Organisation (WHO)

guidelines for particulate matter and level of smoke and dust, a major cause of respiratory diseases. Quality of air in terms of carbon monoxide, sulphur dioxide and nitrogen oxide are in many cases above the WHO standards despite government's action. Slums and squatter settlements are growing at a much higher rate than the urban growth rates stressing the already stressed capacity of municipalities to provide public services. Traffic congestion is another problem, causing air pollution and extending travel times, forcing people to inhale particulate that settles in lungs and thereby affects human health and incurs economic losses.

Cooking Fuels and Indoor Pollution

Burning of biomass and coal for cooking is a major cause of indoor and local air pollution. Vast populations in the slums, in Delhi, are still using fuelwood. The rural population migrating into Delhi continues with their cooking habits. Indraprastha Power Station and Badarpur Power Station are using fossil fuels, especially coal, which are a major source of carbon dioxide as well as sulphur dioxide and nitrogen oxides and cause local and long-range transboundary air pollution. NDMC area has a number of parks and trees. The falling tree leaves are difficult to dispose and therefore, are burnt by gardeners, causing pollution, similar to burning of biomass indoor.

Water Quality

Water quality has been steadily degraded by a combination of factors including human excreta, sewage and industrial effluents, and agriculture runoff. Levels of suspended solids in river Yamuna have increased manifolds since the late 1970s. Water pollution caused by organic matter, pathogenic agents, fertilizers and pesticides, and hazardous and toxic wastes is another serious problem. Bio-chemical oxygen demand has increased and the bacteria load due to human excreta has also increased. Use of inappropriate and outdated technology and lack of financial resources force a number of industries release untreated wastes into Yamuna.

Waste Disposal

Lifestyles based on unsustainable consumption pattern result in waste disposal problem. Rapid rise in income have brought about dramatic increase in solid wastes. The civic authorities — MCD and NDMC — find themselves in a difficult situation to tackle with this problem. Lack of sanitary landfill sites; lack of technology for sewage treatment and lack of financial resources have accentuated the situation. I recall having discussion with a number of NRIs who would virtually have solutions to all problems of sewage disposal. On three occasions I organised meetings with the Chief Secretary of Delhi who was receptive to the suggestions. But, when he asked these NRI professionals to arrange for sewage treatment technology and guarantee performance under conditions prevailing in Delhi, no one came back.

Clean Production Technologies

A major step in improving environmental quality in Delhi has been the implementation of Supreme Court Judgement to shift polluting industries and adoption of cleaner technologies for automobiles and industries. India is one of the countries to have signed International Cleaner Production Declaration (adopted in 1998 in the Republic of Korea) of United Nations Environment Programme (UNEP). But, the effect of such a commitment is still not seen in Delhi.

Environmental Standards

The use of the International Organisation for Standardisation, ISO 14,000 certified environmental management system is becoming more widespread. However, some concerns about the feasibility of small and medium size enterprises adopting these highly structured environmental management systems are being aired. There seems to be a need to promote less complicated and less formulised environmental management system for many of the industries in Delhi. Once this exercise is completed other cities will also benefit from the same.

Information on Environment Management

In order to respond to environmental challenges, access to information is necessary about the issues and management tools such as corporate environmental reporting, environment impact assessment, environmental rating of companies and products, eco-labelling and emergency preparedness plans. Such information includes the environmental performance of individual industrial operations, the potential impact from planned activities, information and environmental characteristics of products and services, and risks and remedial actions associated with industrial operations. There is a need to further adopt and promote the use of a wide range of environmental information and communication tools currently available.

Can't Remain Silent Spectator

It is not appropriate to remain silent spectators to unforeseen growth pattern and be silent spectators. The deteriorating services and environmental situation has thrown 'a challenge' and provided 'an opportunity' to the engineering community to find or evolve solutions to the problem. The situation in the city of Delhi cannot be tackled by the local authorities alone through their own professional, managerial and financial resources. The Delhi State Centre of the Institution of Engineers (India) has to seriously think in terms of the problems of city of Delhi and evolve projects that will lead to solutions through successful demonstrations.

Evolve Networking Arrangement

In its effort to make the city more liveable, the Delhi State Centre needs to take the lead and become nodal point for networking with as large a number of professional and social organisations. We have to search partners — from the Civic Bodies of Government of Delhi and as well from the non-government sectors.

Government's Commitment

I have defined the problems of environmental pollution and in using the assessment tools and awareness creation. To overcome this situation it is best to refer to international strategies.

Commitment at International Level to Fight Environment Pollution

Sustainable development is the key to future global development and quality of life of the people. The Ministerial Conference on Environment and Development in Asia and the Pacific, 2000 held during 31st August and 5th September, 2000 at Kitakyushu, Japan, reviewed major environmental and developmental trends and findings. A Regional Action Programme recognised the overarching need for poverty reduction in developing countries and controlling wasteful consumption patterns in developed countries as fundamental prerequisite for achieving sustainable development. It further recognised that poverty reduction through pro-poor economic growth, social development and the empowerment of people and institutions would contribute to such a progress. The causes and effects of environmental degradation, like those of poverty, are wide-ranging and diverse. Environmental protection can only be effectively realised if it is accompanied by a comprehensive programme for social development and economic growth.

The strategic elements identified by the Regional Action Programme include the following:

(i) Environmental quality and human health;
(ii) Bio-diversity;
(iii) Coastal marine environment;
(iv) Fresh water resources;
(v) Deforestation and land degradation;
(vi) Globalisation and policy integration;
(vii) Climate change; and
(viii) Sustainable energy development.

It can be observed that our thrust on environmental quality and human health and 'natural resource use, such as that of water' is in keeping with the international approach endorsed by the Minister of Environment and Forests of India.

Identified Technological Interventions

While identifying strategic elements, the group also identified the following technological interventions for sustainable growth pattern:

(i) Introduce cleaner production for industries, especially in small and medium-size enterprises, and cleaner technologies for vehicles, and develop cheap and simple technologies for the small-scale sector;
(ii) Encourage the reduction, reuse and recycling of wastes in

industries and the urban domestic sector;

(iii) Encourage integrated pest management and the development and implementation of organic farming techniques, including organic composting techniques for biodegradable wastes, waste minimisation and waste exchange practices;

(iv) Identify a range of environment friendly, socially acceptable and cost-effective technologies and support the transfer of appropriate technologies to the grassroot level through the involvement of local communities, the private sector and other stakeholders;

(v) Provide technical help in creating treatment, storage and disposal facilities for hazardous wastes and chemicals.

Action Programme

Platinum Jubilee

It is a matter of pride for us that during the tenure of our Committee the Delhi State Centre will be completing 75 years of its existence in the year 2002. We will, therefore, like to celebrate the Platinum Jubilee of the Centre in the most befitting manner, focusing on issues that will help in conservation of natural resources and provide better quality of life to the people of Delhi. To celebrate the Platinum Jubilee appropriately we have planned lecture every month in collaboration with prestigious institutions like IIT, Delhi; IARI etc.

Also we plan to organise an International Conference on "Sustainable Growth of Cities in South and South East Asia – Focus on Natural Resources". We had discussions with the United Nations' Social and Economic Commission for Asia and Pacific who agree to support the activity, and arrange participation from countries of the region. We have also had discussion with the International Commission on Irrigation and Drainage and look forward to their partnership in the event.

Projects on Environmental Quality and Human Health in Delhi

It is of utmost importance that the focus of our activities is on the 'environmental quality and human health' and 'efficient utilisation of natural resources'. Of the natural resources, fresh water has been specifically identified as that is considered the major single element in the development process needing maximum attention at the onset of the 21st Century. We would like to identify engineering interventions that will effectively deal with the situation; and then develop implementation mechanism through demonstration.

Since the technological interventions have been professionally accepted and endorsed by the Ministry of Environment and Forest, Government of India, it has eased our task to that extent and has also presented the mandate under which we need to work.

Web Site

We are inaugurating our own web site under the title: www.ieidsc.org. This web site will have the history of the Institution; membership criteria and the current status of members, AMIE examination procedures and all details of the forthcoming events over the next year. Also the monthly newsletter will be on the web for easy access by not only by our members but by all professionals. The members will be informed about the activities of the Centre.

ICID has assured all help and support on a continuous basis.

Virtual Library

The title index, author index, key words and abstracts of books and technical papers will be electronically stored. In due course of time on-line accessing and hyper link to other libraries will be ensured to set up the virtual library. This will also meet the long felt and identified need of our AMIE students.

Transparency, Efficiency and Quality in Functioning

Changing economic and social scenario have thrown open challenge to the engineering fraternity, calling for radical changes in the approach. In this regard the following three aspects are now being focused for action by professional organisations. We would also like to implement them at the Delhi State Centre:

(i) Transparency;
(ii) Efficiency; and
(iii) Quality.

Transparency: Transparency will be brought around, both in our relations with members as well as on the role of our partners. As a first step, we are launching our web site for the Delhi State Centre under the title www.ieidsc.org.

Efficiency: As regards efficiency, we propose to streamline the administrative functioning of our Centre and induct technical personnel to improve the technical services. We shall establish linkage with other professional organisations of national and international repute to augment our technical capability. Some of them have already been identified and we have started discussion. I am glad that the International Commission on Irrigation and Drainage; Indian Institute of Technology, Delhi; and Indian Agricultural Research Institute have agreed to provide all linkages and help and support in our outreach as well as in operations.

Quality: As regards the quality, our aim is to improve the quality of education or guidance to AMIE students by providing them better teaching personnel and linkages with IIT, Delhi and other institutions. We are fortunate to have academicians from all such engineering institutions in our Executive Committee, who have decided to organise such linkages. Our effort will be to ensure recognition of AMIE, wherever it is not recognised so far. This can either be by way of convincing other organisations of the course content or else requesting the headquarters to improve the course contents to be at par with other national engineering institutions.

Quality of service to our members will be improved by sending advance information of the events in terms of calendar for the year and giving them access to the monthly newsletter which will then project the activities in the shorter time frame.

Financing

Delhi State Centre needs to generate its internal resources on a higher scale, which will also be in keeping with the current trend. That will also make our activities more relevant and responsive to the needs of the society. The situation has been alarming. Internal generation of resources can be either through grants from various ministries and donor agencies or else by generating income by conducting specialised courses, training programmes and publications. Financial support from major engineering institutions and industries can go a long way in catalysing such an effort and giving a head-start. We, therefore, request all members to consider themselves and their organisations to donate liberally for maintenance and upgradation of the existing building; creation of modern infrastructure facilities for conferencing/ video-conferencing/tele-conferencing/visual library, and electronic connectivity with IE (I), and national and international organisations. Even small contribution can make a big difference in meeting our target.

Members to be Invited to Executive Committee Meetings

The task ahead is momentous. The Millennium Team of the Executive Committee will need active and dynamic support of all the members, not only 'outside' support but as 'participants in decision making'. We shall endeavour to evolve a mechanism whereby members outside the Executive Committee are invited to its meeting. This will not be an 'arbitrary' or 'ad-hoc' procedure. The invitations will be extended to those who will show their interest. Obviously, everyone cannot be accommodated every time; but, some members can be accommodated sometime. Our sincere effort will be to involve members to their convenience.

Consultative Committee of Past Chairmen and Eminent Engineers

We also propose to set-up a Consultative Committee of all the Past Chairmen of Delhi State Centre which will meet at least once in a quarter, and recommend steps for improving the image and outreach of the Centre. Realising that the Committee will consist of distinguished and experienced engineers of repute, we shall try to get the best advice through synthesis of ideas at brainstorming sessions. We also wish to invite eminent engineers, other than the Past Chairmen, in this Committee.

Fulfilling Obligation to IE(I) Headquarters

The Millennium Team fully realises the responsibility and duties to be discharged to improve the effectiveness of our parent body — The Institution of Engineers (India). We shall try to develop back-up support systems that may support implementation of Council's decisions.

Conclusion

Two years may be too short a time to undertake all the actions to completion. The Millennium Team is determined to make a sound beginning. Maybe when we complete our term each member remembers the famous words of Robert Frost:

'Miles to go before I go to sleep"

The Millennium Team wishes to assure all members that our actions will aim at improving the image of Delhi State Centre; and improving the quality of life of the common man in Delhi, thereby making the Centre an integral part of Delhi's development process.

We look forward to all help and cooperation from our members.

16

Sustainable Development: Outcomes of Johannesburg (WSSD)

Introduction

Efforts to promote Sustainable Development made a significant move forward at the World Summit on Sustainable Development (WSSD), Johannesburg, held during August 26 — September 4, 2002. The Summit concluded with a Statement of Political Commitment to Sustainable Development, backed up by Implementation Plan. The world leaders committed themselves to improve the lives of people living in poverty and reverse the continuing degradation of the global environment. The most significant outcome of the Conference was that agreements were reached on various measures that really could make a difference to the living conditions.

With the positive thinking of development oriented persons the Summit has made a move forward in making Sustainable Development a reality. The global commitments, when backed up by national efforts, will put us on a path that reduces poverty while protecting the environment, a path that works for all people, rich and poor, today and tomorrow.

Success of the Conference must be viewed in terms of the overriding thrust of the Summit: '*to promote action*'. And major progress was made in Johannesburg to address some of the most pressing concerns of poverty and the environment. Commitments were made to increase access to clean water and proper sanitation; to increase access to energy services; to improve health conditions; promote agriculture, particularly in dry lands; and to better protect the world's bio-diversity and ecosystem.

Paper presented at the Seminar on Environment and Development organised by TAFCON on September 4, 2002 at New Delhi.

Rather than concluding with the words and agreed documents, the Summit has also generated concrete '*partnership initiatives*' by and between governments, citizen groups and businesses. These partnerships are bringing with them additional resources and expertise to attain significant results where they matter in communities across the globe.

In the words of Mr. Kofi Annan, Secretary General of the United Nations: "The Summit represents a major leap forward to the development of partnerships with the UN, governments, business and civil society coming together to increase the pool of resources to tackle global problems on a global scale."

As a result of the Summit, governments agreed to a series of commitments in five priority areas that were backed by specific government announcements on programmes and by partnership initiatives. More than 220 partnerships, representing US$ 235 million in resources, were identified during the Summit process to compliment the government commitment, and many more were announced outside of the formal summit proceedings.

Outcome of the Summit

The following is a list of some of the agreements reached and initiatives announced during the Johannesburg Summit. This list is not exhaustive, but reflects some key highlights of the Summit proccss. The commitments shown are those, agreed in the implementation programme adopted at the Summit.

Water and Sanitation

Commitments

- Commitment to halve the proportion of people without access to sanitation by 2015; this matches the goal of halving the proportion of people without access to safe drinking water by 2015.

Initiatives

- The United States announced $970 million in investments over the next three years on water and sanitation projects.

- The European Union announced the 'Water for Life' initiative that seeks to engage partners to meet goals for water and sanitation, primarily in Africa and Central Asia. The Asia Development Bank provided a $5 million grant to UN Habitat and $500 million in fast-track credit for the Water for Asian Cities Programme.
- The UN has received 21 other water and sanitation initiatives with at least $20 million in extra resources.

Energy

Commitment

- Commitment to increase access to modern energy services increase energy efficiency and to increase the use of renewable energy.
- To phase out, where appropriate, energy subsidies.
- To support the NEPAD objective of ensuring access to energy for at least 35 per cent of the African population within 20 years.

Initiatives

- The nine major electricity companies of the E7 signed a range of agreements with the UN to facilitate technical cooperation for sustainable energy projects in developing countries.
- The European Union announced a $700 million partnership initiative on energy and the United States announced that it would invest up to $43 million in 2003.
- The South African Energy Utility Eskom announced a partnership to extend modern energy services to neighbouring countries.
- The UN has received 32 partnership submissions for energy project with at least $ 26 million in resources.

Health

Commitments

- Commitment that by 2020, chemicals should be used and

produced in ways that do not harm human health and the environment.

- To enhance cooperation to reduce air pollution.
- To improve developing countries' access to environmentally sound alternatives to ozone depleting chemicals by 2010.

Initiatives

- United States announced commitment to spend $ 2.3 billion through 2003 on health, some of which was earmarked earlier for the Global Fund.
- The UN has received 16 partnership submissions for health projects with $3 million in resources.

Agriculture

Commitment

- The GEF will consider inclusion of the Convention to Combat Desertification as a focal areas for funding.
- In Africa, development of food security strategies by 2005.

Initiatives

- The United States will invest $90 million in 2003 for sustainable agriculture programmes.
- The UN has received 17 partnership submissions with at least $ 2 million in additional resources.

Biodiversity and Ecosystem Management

Commitments

- Commitment to reduce biodiversity loss by 2010.
- Commitment to reverse the current trend in natural resource degradation.
- Commitment to restore fisheries to their maximum sustainable yield by 2015.

- Commitment to establish a representative network of marine protected areas by 2012.
- Commitment to improve developing countries' access to environmentally-sound alternatives to ozone depleting chemicals by 2010.
- Undertake initiatives by 2004 to implement the Global Programme of Action for the Protection of the Marine Environment from Land Based Sources of Pollution.

Initiatives

- The UN has received 32 partnership initiatives with $100 million in resources.
- The United States has announced $53 million for forests in 2002-2005.

Cross-Cutting Issues

Commitments

- Recognition that opening up access to markets is a key to development for many countries.
- Support the phase out of all forms of export subsidies.
- Commitment to establish a 10-year framework of programmes on sustainable consumption and production.
- Commitment to actively promote corporate responsibility and accountability.
- Commitments to develop and strengthen a range of activities to improve natural disaster preparedness and response.

Initiative

- Agreement to the replenishment of the Global Environment Facility. $2.9 billion.

Conclusion

The Statement on Political Commitment and Plan of Implementation for the World Summit on Sustainable Development provide us with everything needed to make sustainable development take a practical shape over next several years. The test is whether governments, along with civil society and the private sector, can pursue the commitment that are in the documents, and take actions that achieve measurable results. Those, who expected this Conference to find final solutions relating poverty and trade, will certainly feel disappointed.

Note: Statement of Political Commitment and Implementation Plan are given as Annexures.

17

Sustainable Development: The Preparatory Phase of Johannesburg

Introduction

Sustainable Development has been defined in a number of ways over the last decade. All actions as a follow-up of the United Nations Conference on Environment and Development, held in Rio de Janeiro in 1992, have created wider awareness of environment friendly solutions and the need for capacity building in each country. The Rio outcome, as laid down in Agenda-21, provided the *fundamental principles* and the *programme of action* for achieving sustainable development. The United Nations Millennium Declaration and the outcome of the major UN Conferences and International Agreements since 1992 have evolved internationally agreed development goals; and have emphasised on expediting the realisation of the remaining goals of Agenda-21.

The major gap in fulfilment of commitments for implementation of Agenda 21 has been that the benefits could not be provided equitably to all persons. It is now necessary that the World Summit on Sustainable Development should evolve the implementation strategy in such a fashion that the outcomes benefit all, particularly women, youth and vulnerable groups. The implementation should involve all relevant actors through partnerships, especially between governments of the North and the South on one hand, and between governments and major groups on the other, to achieve the widely shared goals of sustainable development. Such partnerships are key to pursuing sustainable development in a globalised world.

The World Summit on Sustainable Development places emphasis

Paper presented at the Workshop on Sustainable Development organised in August, 2002 at Bhubaneshwar.

on Sustainability through optimum interaction of environment and development. 15 years after the term Sustainable Development was coined by the Brundtland Commission as "Development for the present without compromising the development of future generations", the multi-stakeholder global community has moved ahead and emphasised on sustainability. Specific linkages and time bound targets have been assured. It is no longer the decisions of inter-governmental meetings but a synthesis of multi-stakeholder meetings that set the Agenda and decide on the Action Plan and follow-up. Sustainable Development has to be viewed in that light. Present preparatory work for WSSD has clearly indicated the same.

Strategic Areas of Sustainable Development

The Secretary General of the United Nations has outlined five key strategic areas where we must have sense that at the end of Johannesburg Conference we have committed ourselves to action, which goes beyond incremental action — action, which truly marks a quantum change of efforts. The five areas outlined by the Secretary General are (with the acronym WEHAB):

(i) Water,
(ii) Energy,
(iii) Health,
(iv) Agriculture, and
(v) Biodiversity (standing more broadly for eco-system management).

Water

An action programme is to be launched on water to reduce by half the number of people who lack access to clean water by 2015. Presently over one billion people lack access to clean water. It is necessary to support developing countries in developing integrated water resources management and water efficiency plans, by 2005, through appropriate utilisation of river basin, watershed and ground water management as well as introduces measures to improve the efficiency of water infrastructure to reduce losses and increase recycling of water. It will require employment of full range of policy instruments, including regulation, monitoring, voluntary measures, market and information based tools, land use management and cost recovery of water services, and adopting integrated water basin approach. Establishment of public-private partnerships will be equally important.

Energy

An action programme is to be launched to reduce by half the number of people who lack access to modern energy services by 2015. At present, there are more than two billion people in this situation. The Commission on Sustainable Development at its 9th Meeting had focused only on the issue of Energy for Sustainable Development relevant to domestic situations, bearing in mind the principle of common but differentiated responsibilities. It is necessary to integrate energy considerations into socio-economic programmes, especially into policies of major energy consuming sectors, such as the public, transport, industry, agriculture, urban planning and construction sectors. It is also necessary to develop and disseminate renewable energy and advanced fossil fuel technologies with the aim of increasing their share in energy production and consumption, for both domestic and industrial uses. It is also necessary to diversify the energy supply mix by combining, as appropriate, increased use of renewable energy sources, more efficient use of energy, greater reliance on advanced energy technologies including advanced and cleaner fossil fuel technologies and sustainable use of traditional energy resources to meet the growing needs for energy services in the longer term to achieve sustainable development, as well as promote the development of technologies for improving control of the distribution system and encourage further development and implementation of national goals to that end. Improvement in energy efficiency considerations into the planning, operation and maintenance of long lived energy consuming infrastructures, notably transport, urban land use, industry, agriculture and tourism are equally important.

Health

Combating by reducing HIV infection rates by 25 per cent by 2005, in the most affected countries, and globally by 2010 as well as combating Malaria, TB and other diseases is the major target. Common man is entitled to health and protective life in harmony with nature. The goals of sustainable development can only be achieved in the absence of high prevalence of debilitating illnesses while population health requires poverty eradication. There is an urgent need to address the causes of ill health and their impact on development, with particular emphasis on women and children, as well as other vulnerable groups of society, such as people with disabilities, elderly persons and tribals. Strengthening the capacity of health care systems to deliver basic health services to all in an efficient, accessible and affordable manner aimed at preventing, controlling and treating diseases and to reducing environmental health threats is of extreme importance.

Agriculture

Food Security for the common man is a basic right. Over 800 million people in the world suffer from hunger which has to be reduced by half by the year 2015. Agriculture plays a crucial rule in addressing the needs of a growing global population, and is inextricably linked to poverty eradication, especially in developing countries. Sustainable agriculture and rural development is essential to the implementation of an integrated approach to increasing food production and enhancing food security and food safety in an environmentally sustainable way. Various steps are necessary to develop and implement land and water use plans that are based on optimal use of renewable resources and on integrated assessments of socio-economic and environmental potentials and strengthen the capacity of governments, local authorities and communities to monitor the renewable resources. Integrating existing information systems on land use practices by strengthening national research and extension services; and farmer organisations to trigger farmer-to-farmer exchange on good practices such as those related to environmentally sound, low cost technologies with the assistance of relevant international organisations is extremely necessary. Enhancing access to existing markets and developing new markets for value added agricultural products is equally important.

Biodiversity

Biodiversity plays a critical role in overall sustainable development and is essential to our planet and human well being. Convention on Biological Diversity (CBD) is the key instrument for the conservation and sustainable use of biological diversity and to put in place, by 2010, measures to halt biodiversity loss at the global, regional, sub-regional and national levels. It is necessary to incorporate and mainstream the objectives of the Convention on Biological Diversity into global, regional and national, sectoral and cross-sectoral programmes and policies, in particular in the programmes and policies of donor countries and funding agencies. Promotion and support for initiatives of multi-stakeholders is necessary for the conservation of hot-spot areas and promotion of development of national and regional ecological network, in order to conserve and sustainably use biodiversity, recognising the importance of the eco-system approach.

Sustainable Development — Main Elements and Overarching Objectives

Elements of Sustainable Development

The three mutually reinforcing pillars of sustainable development are as follows:

(i) Economic growth,
(ii) Social development, and
(iii) Environmental protection.

Global, regional and national effort is required to promote their integration.

Overarching Objectives of Sustainable Development

The two over-arching objectives essential for attaining sustainable development through promotion of environmental protection:

(i) Poverty eradication; and
(ii) Changing unsustainable patterns of production and consumption.

Sustainability on planet earth can be achieved only when we tackle the issues of poverty and follow the sustainable patterns of production and consumption. This very concept forms the essence for discussion at the World Summit for Social Development. Actions relating to these elements are defined below:

Poverty Eradication

Eradicating poverty is the greatest global challenge facing the world today and an indispensable requirement for sustainable development, particularly for developing countries. While each country has primary responsibility for its own sustainable development and poverty eradication, concerted and concrete measures are required to achieve the internationally agreed poverty related targets. Actions at the international, regional and national levels are required to accomplish these targets. The Millennium Development Goal related to poverty aims at reducing by half, by 2015, of the proportion of people whose income is below US$ 1 PPP per day, and the number of people suffering from hunger, and without access to safe drinking water.

To achieve this target, various international, regional and national actions will include the following:

(i) Establish a World Solidarity Fund for Poverty Eradication and the Promotion of Human Development in the poorest regions of the world, pursuant to modalities to be determined by the UN General Assembly.

(ii) Develop national programme for sustainable development and community development to promote the empowerment of people living in poverty and their organisations. These programmes should reflect their priorities, and enable them to increase access to productive resources, public services and institutions, in particular land, water, employment opportunities, credit, education and health.

(iii) Promote women's participation in decision making at all levels, mainstreaming gender perspectives in all policies and strategies.

(iv) Deliver basic health services for all and reduce environmental health threats, taking into account the linkages between poverty, health and environment, with provision of financial resources, technical assistance and knowledge transfer to developing countries and countries with economies in transition.

(v) Ensure children everywhere, boys and girls alike, will be able to complete a full course of primary schooling and they will have equal access to all levels of education.

(vi) Provide access to other agricultural resources, land tenure arrangement that recognize and protect indigenous and common property resource management systems.

(vii) Build basic rural infrastructure, diversify the economy, improve access to markets and to credit for the rural poor to support sustainable agriculture and rural development.

(viii) Transfer basic sustainable agricultural techniques and knowledge, including natural resource management, to small and medium scale farmers, fishermen and the rural poor, including through multi-stakeholder approaches and public-private partnerships aimed at increasing agricultural production and food security.

(ix) Increase food availability and affordability, including through harvest and food technology and management, as well as equitable and efficient distribution system by promoting community-based partnerships linking urban and rural people enterprises.

Equally important in poverty eradication is strengthening contribution of industrial development and sustainable natural resource management. Such actions will include the following:

(i) Provide assistance to enhance industrial productivity and competitiveness as well as industrial development in developing countries;

(ii) Promote the development of micro, small and medium size enterprises, with a special focus on agro-industry as a provider of livelihoods for rural communities;

(iii) Provide financial and technological support to rural communities of developing countries to enable them to benefit from small-scale mining ventures;

(iv) Provide support to developing countries for the development of low-cost technologies that conserve fuel for cooking and water heating; and

(v) Remove unnecessary regulatory and other obstacles for micro enterprises and the informal sector.

Changing Unsustainable Patterns of Consumption and Production

Fundamental changes in the way societies produce and consume are indispensable for achieving global sustainable development. All countries have to strive to promote sustainable consumption and production patterns, and with all countries benefiting from the process, based on the principle of common, but differentiated responsibilities. Governments, relevant international organisations, the private sector and all major groups should play a critical role in striving to change unsustainable consumption and production patterns. This will basically include the following:

(i) Developing a ten-year work programme for improving resource efficiency in order to promote social and economic development within the carrying capacities of eco systems, while reducing resource degradation.
(ii) Increase investments in cleaner production and eco-efficiency through incentives and support schemes.
(iii) Enhance corporate, environmental and social responsibility and accountability.
(iv) Provide training for relevant authorities at all levels to take sustainability considerations into account in decision-making, including on national investments in infrastructure, business development and public procurement.
(v) Promote the implementation of recommendations that relate to energy for sustainable development, taking into account that energy is central to achieving the goals of sustainable development.
(vi) Promote an integrated approach to policy making at national and regional levels for transport services and systems to promote sustainable development, including policies and planning for land use, infrastructure, public transport systems and goods delivery networks, with a view to providing efficient transportation, reducing energy consumption and pollution, reducing congestion, limiting urban sprawl, and promoting long term sustainable development taking into account national priorities and circumstances.
(vii) Prevent and minimize, waste and maximize reuse and recycling, with participation of government and all stakeholders, in order to improve resource efficiency.

Governance for Sustainable Development

The potential of globalisation to promote *sustainable development for all* still remains a distant dream. The Secretary General of the United Nations in his Millennium Report had warned that unless the globalisation process brings the poor centre stage it will lose its meaning and effectiveness. As globalisation has been accepted the route to poverty reduction and sustainable development, it is necessary to ensure governance. Good governance practices are to evolve to ensure national economic and social policies in such a fashion so that sustainable development becomes a feasible option. Governments need to announce policies that lead to wealth creation, that in turn will support social development.

The Committee on Constitutional Reforms in India under Chairmanship of Justice M.N. Venkatchaliah, the former Chief Justice of India, has defined governance to include actions of governments, corporates and the civil society. Any discussion on good governance will have to focus on functioning and actions of these three groups distinctly and not of only one of them. Good governance within each country and at the international level is essential for sustainable development. Unfortunately, corruption has been projected as the legacy of developing world and is reflected to be a sign of poor governance. Very little was known about the high level of corruption practiced in developed countries under the garb of efficient operational mechanism of large corporates. However, the ENRON debacle has brought forth that corporates all over the world have been evolving financial norms and procedures that did not reveal the correct picture to its shareholders and stakeholders. Maybe by the time the World Summit on Social Development will take place, the issue of corruption will be linked to the way the corporates in developed countries have embezzled large sums of common man's money. This is not to say that just because corruption takes place in the North, it can be tolerated or pursued in the South. Good governance which included good corporate governance and effective civil society actions — will remain the most important element for sustainable development.

At the domestic level, sound environmental, social and economic policies, democratic institutions responsive to the needs of the people, rule of law, anti-corruption measures, gender equity and enabling environment for investment are the basis for sustainable development. As a result of globalisation, external factors have become critical in determining the success or failure of developing countries in their national efforts. The gap between developed and developing countries point to the continued need for dynamic and enabling international economic environment supportive of international cooperation, particularly in the field of finance, technology transfer, debt and trade, full and effective participation of developing countries in global decision making, if the momentum for global progress towards sustainable development is to be maintained and increased.

Some of the actions required include:

(i) Adopt and implementation of coherent and sound micro-economic polices and development of institutional capacities.

(ii) Create and open, equitable, rule based, predictable and non-discriminatory multilateral trading and financial system that benefits all countries in the pursuit of sustainable development.

(iii) Make foreign direct investment more supportive of sustainable development and encourage international and regional institutions as well as institutions in some countries to promote increasing investment flows to developing countries and to assist developing countries in their effort to create a conducive, domestic and enabling environment.

(iv) Promote corporate responsibility and accountability and the exchange of good practices including through public-private partnerships and voluntary initiatives building, *inter alia*, on the UN Global Compact and the OECD Guidelines for Multilateral Enterprise, as appropriate, as well as by using tools such as Environmental Management Accounting and Environmental Reporting.

Enhancing corporate environmental and social responsibility and accountability will include the following elements:

(a) Encourage industry improve social and environmental performance through voluntary initiatives, including environmental management systems, codes of conduct, certification, and public reporting on environmental and social issues, taking into account such initiatives as the International Organisation for Standardisation (ISO) standards and the Global Reporting Initiative (GRI) guidelines on sustainability reporting.

(b) Encourage dialogue between enterprises and the communities in which they operate; and other stakeholders.
(c) Encourage financial institutions to incorporate sustainability considerations into their decision-making processes.
(d) Develop workplace-based partnerships and programmes, including training and education programmes.

Preparatory Committee IV Deliberations

The Inter-Governmental Preparatory Committee IV, popularly called PrepCom IV has concluded at Bali on June 7, 2002. This meeting has outlined the agenda and the documents for Declaration that will have to be discussed and endorsed at the Johannesburg Conference to ensure the process of Sustainable Development within the framework of the Millennium Targets defined in the Report of the Secretary General of the United Nations. A number of important issues have remained undecided.

Mr. Nitin Desai, Assistant Secretary General, UN, and the Secretary General of the World Summit on Sustainable Development, in his opening remarks said:

> "The main issue is not only closer negotiations on the programme of action, the issue is also one of whether what we come out with is bold and firm enough to meet the high aspirations that people have for the Johannesburg Summit."

Responding to this call the world environment leaders realised the challenge is not just agreeing, but also of agreeing on something that the world recognises as being as major step forward for Sustainable Development.

These challenges can be summarised as follows:

(i) Many people see the Johannesburg Summit as part of a set of Conferences that are defining a new *multilateralism*. The Millennium Summit was a major step in the series of UN Conferences held in the 90s. In Doha (WTO Conference) a major step forward was observed in the way in which at least some of the concerns of developing countries were given a central place in the World Trade Agenda. This is perhaps the first time that we will have trade round whose primary focus is going to be the concerns of developing countries.

(ii) The Global Conference on Finance for Development, in Monterrey, sought to do the samething for the world of finance — to put development at the centre of the world of global finance. In large measure a consensus was arrived at Monterrey. Even more than that, Monterrey was marked by major commitments of additional resources by the European Union, the United States and others, reversing what has been nearly a decade of decline in ODA; and reversing it in a very substantial fashion with a total increase in assistance pledged—that could amount to as much as 12 to 13 billion dollars extra by 2006.

(iii) Thus, the Johannesburg Summit becomes the third leg of this exercise, where the leaders have to define how sustainability can be put into development in order to give new multilaterlism a pragmatic basis for cooperation.

The major challenge, therefore, is to ensure that what comes out of Johannesburg is seen as a step forward in outlining a new programmatic basis for development cooperation drawing on the principles that were set out in Rio and subsequently.

It means, that the Johannesburg Conference, as we all have accepted, is not about renegotiating policy frameworks. The real challenge at this conference is to see how we can put a commitment to credible action in what we negotiate. The focus is *on action*, and *on implementation*, which is universally accepted by all participants, to have clarity about ends and means.

New monitoring and reporting mechanisms are also essential to ensure responsibility and accountability. This is going to be their measure of success of the extent to which the programme will be discussed at the Johannesburg Summit.

Major Issues in Future Actions

Three major issues were considered by the PrepCom IV:

(i) Sustainable Development Governance
(ii) Capacity Building for Sustainable Development
(iii) Major Groups' Framework for Partnership Initiative

The PrepCom IV included a multi-stakeholder dialogue segment during 27-29 May, 2002 involving government and all nine major groups of Agenda-21. The different stakeholders expressed their opinions differently highlighting their thrust areas. These summed up issuewise as follows:

Sustainable Development Governance

Different groups expressed their opinions as follows:

(i) *Women* recalled the vision from Rio linking economic, social and environmental sectors involving common but differentiated responsibility; and stated that this vision requires strong global, national and sub-national institutions with the necessary mandate, authority and resources. They asked for more participation in decision making, and called for government action in priority areas of globalisation, gender justice, transparency and accountability.

(ii) *NGOs* addressed participatory right in decision-making, the role of financial and trade institutions and corporate governance. Referring to principle 10 of Rio Declaration they said good governance demands public participation, access to information and access to justice. They noted that decisions by the WTO and international financial institutions have accelerated unsustainable practices, debt, and supported use of the precautionary principle in international field.

(iii) *Trade Unions* highlighted workers' participation and proposed conducting work place assessments over the next decade to access progress and support, workers participation in seeking joint solutions and challenged employers and governments to make this happen.

(iv) *Local authorities* noted the success of initiative since Rio and stressed international governments should acknowledge the role they could play in assisting national Agenda-21 implementation. They emphasised that local government is not non-government, but the sphere of government closer to the people and thus suitable implementation initiative.

(v) *Business and industry* welcomed the combination of

Type-I and Type II outcomes, stressing inclusion of stakeholders as essential for the success of the emerging sustainable development partnership paradigm. They supported transparency and the global reporting initiatives in this context.

(vi) *Scientific and technological communities* emphasised the need to improve collaboration between scientists and policy makers. They supported dialogue at all levels to ensure policy relevance, education in science and technology for sustainable development, capacity building to bridge the North-South divide, access to information including remote sensing for monitoring and proposed an Advisory Panel on Commission for Sustainable Development on Science and Technology.

(vii) *Farmers* noted the need to ensure food security and safety and called for policies addressing domestic markets, rural development, and better access to and management of natural resources. They underscored greater involvement of farmers at all levels of decision-making and stressed that increasing trade is good only if it enhances prosperity for rural people. Youth stressed the promise of Rio for inter-generational equity and called for more action-oriented outcome from WSSD while noting that the power of MNCs may have undue influence on this process.

During this session, one or more Major Groups recommended that governments:

- Adopt a legally binding convention on corporate accountability, with independent mechanisms for monitoring, compliance, enforcement, and liability;
- Achieve equal representation of women in political, social and economic decision-making at all levels;
- Address the lack of institutional democracy in the World Bank, IMF and WTO;
- Seek transparency and accountability from multilateral and international donor agencies and fund allocation and implementation of project in recipient countries or through agencies.

- Include all stakeholders in national sustainable development strategies;
- Adopt a "rights-based" approach to community access and control over natural resources;
- Conclude the Decade of Indigenous People with a world conference on indigenous people and sustainable development, and adoption of a universal Declaration on the Rights of Indigenous People;
- Utilize the Permanent Forum on Indigenous Issues as a focal point for sustainable development partnerships and implementation plans;
- Recognize the primary role of local authorities in sustainable development governance;
- Employ local skills and introduce new technologies locally to further poverty eradication;
- Encourage initiatives that promote synergies between ODA, FDI, and good governance;
- Establish a CSD advisory panel on science and technology;
- Launch a research and development initiative focusing on interactions and interdependencies between social, economic and environmental systems;
- Develop commonly accepted indicators at the national level;
- Assess impacts of international trade agreements on sustainable development;
- Develop partnerships to monitor and control trade and promote policies that address the needs of producer countries;
- Eliminate corruption through initiatives to strengthen local government;
- Ensure that MEA's are not subordinate to or undermined by WTO agreement; and
- Establish national youth councils to ensure youth engagement in decision-making.

Capacity Building for Sustainable Development

Different groups expressed their opinions as follows:

(i) *Women* stressed Sustainable Development education as a

priority, called for capacity building for gender equality and participation in disseminating, and proposed capacity building measures such as dissemination of gender disaggregated data at the establishment of grassroots women's academies. They indicated the need for commitment of resources for gender mainstreaming and gender budgeting at all levels. Youth stressed holistic formal education that promote sustainable development principles and called for establishment of information clearing houses at the sub-regional level (by 2005) funded by national and local authorities.

(ii) *Tribals* (indigenous people) pointed out that the emphasis of the current development model on marketing rather than conserving natural resources is in appropriate and not complementary of Sustainable Development principles. They supported strengthening existing indigenous capacity on the basis of traditional knowledge and indigenous institutions, complimentarity between scientific and indigenous knowledge, and technology transfer that respects indigenous identities.

(iii) *NGOs* focused on participatory decision making processes for capacity building and advocated empowerment and decentralisation of resources and responsibilities. They considered the principle of common but differentiated responsibilities, crucial for equitable partnerships and raised access to information, modern technology, and finance as essential components for capacity building.

(iv) *Local authorities* stressed that building their capacity will improve implementation of sustainable development. They emphasised the need for national level support to strengthen local capacity, competence and resources.

(v) *Business and industry* pointed to the diversity within their sector and that large and small businesses have different contributions in sustainable development at different levels. They said, business is an important vehicle for capacity building in developing countries through domestic and international transfer of and investments in knowledge, skills, finance and technology. Good governance, the rule of law, and increased funding for scientific research are important to support these ongoing contributions.

(vi) *Farmers* called for partnership in legal framework that enables income-generating activities. They called upon governments to provide incentives for sustainable agricultural practices, and to ensure a balance between local and export markets.

(vii) *Initial government* responses were made by EU, Brazil, Sweden, India, Canada, China and Venezuela. India outlined the need to develop inventories that disseminate, replicate and further innovate best practices from developing countries to encourage South-South cooperation.

The following critical elements for capacity building emerged from the dialogue:

- Good governance with participatory, transparent and democratic decision-making mechanisms;
- Partnerships between states and stakeholders in decision-making processes that focus on building mutual capacities;
- Access to basic services for vital human needs, education, information, modern technology and finance;
- Rights based approach to sustainable development;
- Prior informed consent in the design and implementation of strategies;
- Science-based decision making;
- Genuine partnerships with a level playing field with local ownership of the process;
- Common but differentiated responsibilities;
- Workers rights, freedom of association and collective bargaining; and
- Respect for cultural diversity.

The following means for capacity building were suggested:

- Education for sustainable development and the eradication of illiteracy;
- Dissemination of information (in local languages, gender-disaggregated);
- Access to information technology and related training;

- Decentralisation of resources and responsibilities;
- Enabling environment for employment and income generation;
- Building and maintaining scientific capacity in all countries;
- Trade related technological assistance; and
- Strengthened Local Authorities.

Participants also made the following recommendations:

- Create grassroot academies and peer-to-peer learning institutions;
- Develop gender budgeting, monitoring and resources for gender mainstreaming;
- Establish national youth development programmes for youth participation;
- Enable complementary between indigenous knowledge and scientific knowledge;
- Strengthen indigenous institutions for promoting knowledge sharing;
- Foster local leadership;
- Enforce and implement existing legislation;
- Recognize business as an important vehicle for capacity building;
- Ensure balance between local and export markets and expand market access for developing countries;
- Support, promote and finance Local Agenda 21 activities and campaigns;
- Develop inventories to disseminate, replicate and further best practices;
- Share environment friendly technology across countries through an international mechanism; and
- Prevent brain drain and build domestic capacity by providing incentives, financing and training for citizens of developing countries.

Major Groups' Framework for Partnership Initiatives

(i) *Women, youth, NGOs and tribals* stressed that the outcome should not result in unequal power relations, support unsustainable economic development models, or inappropriately accelerate the expansion of the private sector's role in providing development services. They also urged strong commitments for corporate accountability and ratification of existing conventions.

(ii) *NGOs* expressed concern over the increasing influence of Corporate Power in the UN and its agencies, stressed the need to ensure that Type II initiatives are anchored in Type I commitments and proposed that partnerships involved only those governments that have ratified multi-lateral environmental agreements. They stipulated that Type II initiative be transparent, consultative, protect the rights of individuals, not exclude legally enforceable contracts among partners and have credible monitory mechanism to allay fears of those who mistrust Type II partnerships.

(iii) *Local authorities* stressed monitory and evaluation of Type II partnerships as a means to implement targets, noting that partnerships are not an end in themselves. They called for partnerships that are action oriented, inclusive, transparent, and not a replacement for institutional responsibilities. They noted the importance of strengthening local authorities as a bridge between local and international levels of implementation and monitoring.

Conclusion

Sustainable development is probably the most daunting challenge that humanity has ever faced, and achieving it requires that the fundamental issues be addressed immediately at local, regional and global levels. Bridging the development gap between the North and the South, and alleviating poverty to provide a more equitable and sustainable 'future for all', requires novel integrated approaches that fully incorporate existing and new scientific knowledge. Whatever the cultural, geographic, socio-economic and environmental setting, a strong partnership between the S&T community and other members of civil society, the private sector (or the corporates) and governments is a fundamental prerequisite to sustainable development.

The Johannesburg Summit is not called to endorse business as usual in public policy or private activity. The Summit has been called because people want change in public policy and private activity: it is necessary to develop a credible mechanism for ensuring that these changes will take place in the years beyond Johannesburg Summit.

The expectation from the WSSD can be best summed up by a quotation from the document of the Worldwide Fund for Nature. It reflects the sentiments and concerns of many of the non-governmental organisations. A passage reads as such:

> "The Summit will be a failure if government leadership is not shown in the form of strong action plan with targets and timetables, and commitment to the resources needed to support implementation of the action plan."

18

Engineers and Sustainable Development

The United Nations Conference on Environment and Development, held in Rio de Janeiro in 1992, provided the fundamental principles and the programme of action for achieving sustainable development. The leaders from different countries have strongly re-affirmed commitment to the Rio principles, the full implementation of Agenda-21 and the programme for the further implementation of Agenda-21. Other major international conferences and the United Nations Millennium Declaration have supported and identified development goals.

To expedite the realisation of the remaining goals all stakeholders have to undertake concrete actions and measures at all levels, bearing in mind the principle of common but differentiated responsibilities. These efforts will have to promote the integration of the three components of sustainable development: economic growth, social development and environmental protection, as mutually reinforcing pillars. Poverty eradication and changing unsustainable patterns of production and consumption are overarching objectives of sustainable development and an essential requirement of promoting environmental protection.

Good governance, which includes corporate governance, within each country and at the international level, is essential for sustainable development. At the domestic level, sound environmental, social and economic policies, democratic institutions responsive to the needs of the people, rule of law, anti-corruption measures, gender equity and enabling environment for investment are the basis for sustainable development. As a result of globalisation, external factors have become

---

Welcome address at the Round Table Discussion on the Text for Negotiation at the Preparatory Committee for the World Summit for Sustainable Development organised by IEI and ICID on May 31, 2002 at New Delhi.

critical in determining the success or failure of developing countries in their national efforts. The gap between developed and developing countries point to the continued need for a dynamic and enabling international economic environment supportive of international cooperation, particularly in the field of finance, technology transfer, debt and trade. Full and effective participation of developing countries in global decision-making, if the momentum for global progress towards sustainable development is to be maintained and increased.

Sustainable development requires inputs from science and technology and engineering. Scientific knowledge and appropriate technologies are central to resolving economic, social and environmental problems that make current development process unsustainable. Bridging the development gap, between the North and the South, and alleviating poverty to provide a more equitable and sustainable future for all, requires innovative integrated approaches that fully incorporate existing and new scientific knowledge. The S&T community is expected to make a leading contribution to tackling major problems identified in Sections III & IV of the Millennium Declaration — 'Freedom from want' and 'Sustaining our future'.

WFEO has defined Engineering as the art of using science and technology, traditional experience and creativity, for the benefit of humanity. The engineering profession is a major contributor to the quality of life, welfare and wealth of the people of the world. Most social and economic activities depend increasingly on engineering applications. Engineering skills provide housing, civil structures and infrastructure, services for storage and distribution of water and other services vital to community health, for the generation, transmission and use of power, for communications, and for production of goods and services including transport and processing of food. Increasingly the standards of engineering knowledge and engineering practice determine the long-term welfare of people and the wealth generation capabilities of nations. In recent years it has become evident that particular engineering skills are required to reduce demand on natural resources, and to restore and safeguard the bio-physical environment in which we live.

As population and economic growth place increasing pressures on our social and bio-physical environment, engineers must accept increased responsibilities to develop sustainable solutions to meet community needs, overcome extreme poverty and prevent segregation of people. The education of engineers needs to inculcate an understanding of sustainability and cultural and social sensitivities as well. The engineering code of ethics must reflect a strong commitment to principles of sustainable development.

To achieve the Millennium Development targets, there is an urgent need for partnerships between scientific, engineering, public and private institutions; and integration of scientists' and engineers' advice into decision making bodies in order to ensure a greater role for science, technology development and engineering sectors.

Implementing Agenda-21 during the next decade and beyond will build on progress made during the past ten years in generating knowledge targeted to sustainable development objectives and in developing cleaner and more affordable technologies. However, enhancing the S&T and engineering community's capacity to contribute towards sustainable development will require significant actions. This includes the following:

(i) More policy relevant science, technology and engineering,
(ii) Broad based participatory approaches,
(iii) Promoting gender equality in science.

The present global S&T system needs to be re-directed towards sustainable development challenge. Following are crucial components of such a new arrangement between S&T and Society for Sustainable Development:

(i) Poverty alleviation, food and water security, and better health,
(ii) Science and Technology for Sustainable Development,
(iii) Improving education and capacity building,
(iv) Clean technologies and sustainable production and consumption patterns,
(v) Lending traditional and scientific knowledge for sustainable development,
(vi) Bridging the North-South divide in Scientific and technological capacity,
(vii) Governance for Sustainable Development.

Two documents are under serious discussion at the PrepCom IV at Bali, Indonesia. These are as follows:

(i) Chairman's text for Negotiation.
(ii) Dialogue Paper by Scientific and Technological Communities — Science and Technology as a Foundation for Sustainable Development — for the Multi-Stakeholder Dialogue Segment.

The first Document is the Draft for Political Commitment. The Draft has specified engineering inputs for the sustainable development and its crucial role. Several of the clauses define this role.

The second document has collaboratively prepared by ICSU and WFEO. The role of scientific community is highlighted and engineering aspects are mentioned only as a part of technology development.

19

Sustainable Development: Overview of Issues

Introduction

The United Nations Conference on Human Environment, held at Stockholm in June 1972, was an eye opener to the developed as well as developing countries to discover the fragilities of our natural systems. The growing phenomena of environmental degradation and depletion of industrial resources were considered at length by the governments of the world. Poverty was projected as a major cause for environmental degradation. Poverty embraced the lack of capacity to buy services that would ensure cleaner development or can result in cleaning up the degradation effects. The then Prime Minister of India projected the problem of the developing countries very cogently in the Stockholm conference. It was emphasised that without poverty alleviation, pollution abatement would not be feasible. Mrs. Indira Gandhi, the Prime Minister of India, told the world that 'poverty is the worst polluter'. Twenty years later, the United Nations Conference on Environment and Development (UNCED), held at Rio de Janeiro in June 1992, evolved the Agenda-21 for human development in an environment-friendly manner. This long journey was from 'awareness creation' to 'action programme'. The action programme aims at preparing the world for the challenges of the next century. At the centre of this 'action programme' is the spirit of self-supporting, 'people-led' mechanism leading to sustainable development. An important message was that 'faster economic development in the developing

---

Paper presented at World Congress on Sustainable Development, organised by the Institution of Engineers (India), under the aegis of World Federation of Engineering Organisations during January 20-23, 2000 at Calcutta.

countries can be achieved only through industrial growth in harmony with the environment'.

The present patterns of development poses serious threat to the global eco-system. The poor are often blamed for living off the natural resources of the land and their environment, and consumption patterns are often held responsible for ever-increasing material comforts of the economically affluent which leads to severe stress on the environment. While the consumption patterns of the poor are unsupportable in terms of land use and cause degradation of the village commons, the adverse impact of lifestyles of the rich affect the global commons. It is easy to understand the environmental degradation by the poor in their fight for survival. But, the rich with no such compulsions have really to control their consumption patterns if the global environment is to be preserved. It is extremely necessary to focus attention on the development that is human-centred and supports growth of goods and services needed for improving the welfare of mankind. Thus two different scenarios control the world:

(i) Poverty in developing countries.
(ii) Consumption patterns in developed countries.

The concept of sustainable development can really be implemented through a balance of 'need' (of poor) and 'greed' (of rich).

Global Overview

The UNEP report, 'Global Environment Outlook-1' (1997) mentions that significant progress has been made in the last decade in confronting environmental challenges in both — the developing and the industrial regions. Worldwide, the greatest progress has been achieved in the realm of institutional developments, international cooperation, public participation, and the emergence of private sector action, legal frameworks, economic instruments, environmentally sound technology, and cleaner production processes. Environmental impact assessment has become standard tool for the initiation, implementation, and evaluation of major development and investment projects in countries.

As a result, several countries report marked progress in curbing environmental pollution and slowing the rate of resource degradation as well as reducing the intensity of resource use. The rate of environmental degradation in several developing countries has been slower than that experienced in industrial countries when they were at a similar stage of economic development.

Despite this progress on several fronts, from a global perspective, the environment has continued to degrade during the past decade, and significant environmental problems remain deeply embedded in the socio-economic fabric of nations in all regions.

Although poverty and the growing global population are often targeted as responsible for much of the degradation of the world's resources, other factors—such as the inefficient use of natural resources, waste generation, industrial pollution and wasteful consumption patterns—are equally driving us towards an environmental precipice. In many instances, although environmental degradation trends are rising, the rate of increase over the years has slowed down comparable economic growth. This suggests that several countries are making the transition to a more sustainable environment at a lower level of economic development than industrial countries typically did over the last 50 years.

Several issues that need to be considered for sustainable development include the following.

Population

Most of today's problems are related to uncontrolled human populations exerting pressure on natural resources. UNDP has projected that the pressure has been escalating rapidly, to a large extent as a result of the dramatic increase in the world population over the last century. It is now clear that the tremendous increase in the recent past in the world's population is associated with what is known as a demographic transit. In such a transition, the population moves from an old balance characterised by 'high mortality and high fertility rates' to 'low-mortality and low-fertility rates'. Demographic transitions have occurred in the past, in Western Europe in the 19th Century, and in Southern and Eastern Europe in the first quarter of the century. They are now taking place in all the developing countries.

The demographic transition currently observed in developing countries has been initiated by the rapid decline in mortality in these countries brought about by improvements in public health and advances in medical technology. For example, an increase in life expectancy from 40 to 50 years was achieved in developing countries in only 15 years from 1950-65. If a large reduction in mortality rates is not accompanied by a fall in fertility rates, the population will increase indefinitely.

This population momentum has important geographical, locational and age dimensions as follows:

(i) Uneven geographic distribution of population growth: 90 per cent of the additions to the world's populations are taking place in the developing countries. (Current additions to population are primarily in people abundant countries with low average incomes.)

(ii) Locational distribution of population growth is such that the urban share of population growth has increased and will continue to do so.

(iii) Age distribution of the population is changing in all countries, but the nature of change is different across countries. (The population is becoming older in the rich countries because life expectancy is increasing, and where infant mortality is relatively stable. In the poor countries, the population is becoming younger, because infant mortality is declining much more than the increase in life expectancy).

The population trends in developing countries are such that the percentage of population in urban areas is increasing. Thus the problems will be gradually shifted from rural to urban areas. Most of these populations will shift to urban centres without improvement in infrastructure, thereby over-burdening the existing infrastructure beyond its carrying capacity. Moreover, since the urban slums will attract government's attention much more easily, the developmental public funds will be diverted to the urban areas at a faster pace (at the cost of rural areas).

Water

UNEP has projected that everyday, 25,000 people die as a result of poor water quality. Waterborne diseases still represent the single largest cause of human sickness and death worldwide. Some 1.7 billion people, more than one-third of the world's population, are without safe drinking water supply. In addition, an estimated one-quarter of the world will suffer from chronic water shortages in the beginning of the next century.

Water supply to regions hosting mega cities is a concern worldwide, particularly with regard to protection of ground water resources, intrusion of salt into fresh water supplies and land subsidence. More than 1.5 billion people depend on ground water for drinking. Other global priorities are the equitable distribution of water among riparian countries, sharing international river basins, non-point sources of pollution and the impacts of major dams and diversion projects. Water will be the major impediment to development in the future of several regions. Water is likely to be that the biggest challenge of the next century.

About 60 per cent of the world's population lives within 100 kms of coastlines and more than 3 billion people rely in some manner on coastal and marine habitats for food, building sites, transportation, recreation and waste disposal. Around one-third of world's coastal regions are at high risk of degradation, particularly from land based sources of pollution and infrastructure development. European coasts are most effected with some 80 per cent at risk, followed by the Asia and Pacific with 70 per cent of the coast at risk.

India faces the basic problem of safe drinking water for the rural, and now also for the urban populations. Appropriate use of surface water through large dams has also become a difficult proposition. Implementation of Sardar Sarovar Dam Project has become a challenge. Environmental activists have more or less sealed the fate of large dams.

Air

Air pollution problems are multi-faceted and all pervasive, that have a direct adverse effect on sustainable development. All major cities in the world suffer urban air quality problems. In Eastern Europe air quality is considered the most serious environmental problem. Acid rain and trans-boundary air pollution, once problem only in Europe and parts of North America are now becoming apparent in parts of the Asia-Pacific region as well as parts of Latin America. Large regions are at risk from the effects of both climate change and acidification. Despite coordinated action worldwide, damage to the ozone layer continues faster than expected, with the next 10 years predicted to be the most vulnerable. All regions expressed concern over global warming, but special emphasis is placed by developing countries on the need for adaptive mechanisms to cope with accompanying climate variability and sea-level change. The rapidly rising demand for energy to fuel economic development is expected to aggravate these problems, particularly in the Asia and the Pacific, where a 10 per cent increase in energy use is predicted for 1990-2010, and in Latin America, where 55.77 per cent increase of energy is predicted. It is expected that for the near future, fossil fuels will continue to be the primary energy sources. The Kyoto Protocol recommends major steps to contain these changes. The outcome of the Third Conference of the Parties to the United Nations Framework Convention on Climate Change, held in Kyoto, Japan in December 1997, was also disappointing in many respects. Industrialised countries agreed to reduce their greenhouse gas emissions by an overall 5.2 per cent from 1990 levels by 2008-2012, but this was less than the European Union (EU) and some governments proposed originally. However, the agreement does cover six gases while initially only the carbon dioxide, methane and nitrous oxide were to be covered. Although the Kyoto Protocol does not detail specific policies and measures for reducing greenhouse gas emissions, it does say that 'each industrialised country has to implement and/or further elaborate policies and measures in accordance with his national circumstances'.

The Kyoto Protocol is open for ratification by the member countries till end of 2000. The Clean Development Mechanism which will 'decide pace of action' under the Protocol is to be adopted by end January 2000. Only a handful of countries have ratified the Kyoto Protocol by end December 1999. The carbon trading mechanism is by and large hotly debated and there is little chance of its ratification within this stipulated time.

Role of Agriculture in Fighting Hunger and Malnutrition

'Some 840 million people go hungry everyday and at least 1.3 billion live on less than US$ 1 per day. If the world can't make progress against hunger and poverty, by year 2025, there could be 4 billion people living on less than US$ 2 per day and more than 2 billion living in extreme poverty', says the World Bank. Each year the global population climbs by an estimated 90 million people. This means, at the very least, the world's farmers will have to increase food production by 50 per cent to feed some 2 billion more people by 2020, and may be as much as 100 per cent if current trends in increased meat production — and the accompanying demand for more foodgrains — continue. These stark numbers have led some analysts to believe that what is needed is a new and greener revolution to once again increase productivity and boost production.

The statistics on food security appear grim. In addition to the expected population growth, the World Bank as well as the FAO estimated in 1996 "as many as 840 million people — a number that exceeds the combined populations of Europe, United States, Canada and Japan — currently do not have enough to eat." (In the year 2000 this figure has been down rated to 824 million). That means one-fifth of the people living in developing world do not get enough calories, enough protein or both. The resulting malnutrition is particularly harmful for children and pregnant women. According to WHO, an estimated 19,000 infants and children are among the 40,000 people who die each day of problems linked to malnutrition.

The companion problem of hidden hunger — deficiencies of vital micro nutrients — affects even more people in the developing world. An estimated 2 billion lacks, sufficient iron in their diets, with about 1.2 billion weakened by iron deficiency anemia. Vitamin-A deficiency affects about 125 million school children and has produced irreversible eye damage in an estimated 14 million; another 600 million people suffer iodine deficiency disorder.

Hunger and malnutrition of this dimension has been projected among the rural poor, who could not grow enough food to meet their needs. Now, it has spread to growing numbers of the urban poor — many of whom have left the rural homes and moved to cities — who are not able to afford the food they need. Often food is available, but not accessible.

But the numbers don't tell the full story. The challenge confronting the world is far more complex than simply producing more food because global conditions are very different than they were at the time of green revolution. Easy gains in food crop yields seemed to have been achieved in some regions, yields may have reached their ceiling. Environment issues have gained importance. Many of the most promising lands are already under cultivation, erosion is taking a growing toll, water shortages loom in many areas, and the majority of the world's fisheries are overexploited.

India has made significant improvement in cereal production, reaching 200 mt in 1999, thus keeping pace with increasing population. Cereal availability in the range of 475-500 gms per capita has been maintained. India has also overcome the acute hunger, or 4th degree hunger, but not malnutrition.

Poverty

Throughout the developing world, poverty is linked to hunger. In Sub-Saharan Africa, plagued by malnutrition, every other person is poor. There and in South Asia — another badly under-nourished region — more than half of the poor live in rural areas, where lands are marginal and their livelihoods depend mainly on agriculture.

Poverty is the bane of developing countries. Poverty in the developing countries has many reasons including political and social discrimination. But many are poor, because they have no tangible assets, no land and no livestock, and lack formal education and technical skills. Many people either settled where the land is only marginally productive best and where governments have failed to provide the basic infrastructure, essential to economic development, or they migrated to urban areas.

Rural poverty and accompanying malnutrition are usually tied to land. Much of the land owned by poor families has been divided and subdivided so often that what remains cannot support a family. The general impression, emerging out of the developed economies has been that a market based economic development is the ultimate solution for all problems of poor. However, the emerging trends indicate that lack of purchasing capacity, deprives the poor to have any role or even access the market based interventions. Thus they become 'market excluded'. If they are market excluded, then how can they take advantage of economies of scale based on global market competitiveness. All international agencies, including the World Bank and the International Monetary Fund (IMF), have realised that poverty reduction cannot be achieved through market interventions alone. Agriculture and rural development are, therefore, considered central to the poverty reduction mission.

India has made significant strides in overcoming poverty; but all such planned actions get nullified because of unplanned population growth. Over 300 million rural people are reported to be below poverty line — a level decided by the government.

Unemployment

Unemployment is an urgent problem even in modern industrial societies. The market economy measures lead to believe that highest productivity per person (and not that highest employment) holds the key to development.

The term "work" was not restricted to paid employment until the industrial revolution. It is difficult to say if a new definition of work will emerge in the near future. Poverty reduction is fully related to creation of income generation opportunities either through self-employment or through employment opportunities. Policies aiming at a return to a classical growth and transformation into the service society fail to recognise as to what has been experienced in developed countries for sometime now can no longer be explained by a cyclical economic crisis, nor by the problems of adjusting globalisation. The fact is that the fast growing industrial economies are facing the worst challenge due to jobless growth. They are witnessing an end to the era of European paid labour society with full time work. Whole range of factors — the end of the long post-war recovery, the migration of investment (caused by globalisation) and above all the micro-electronic revolution — are making it impossible to retuın to classical full employment, even if the developed countries manage to achieve high growth rates. Moreover, for ecological reasons, classical growth would in fact be very problematic. Innovation driven growth in cutting edge technologies might boost the national product, but it will not increase many jobs.

It is, therefore, necessary to think about full employment and distribution of work. That will really and truly represent sustainable growth.

The development of incomes since 1992, in India (in the post globalisation period) have shown that wage restraint does not create jobs. Making working hours more flexible, reducing the amount of overtime work and coupling this to work sequences can overcome cyclical problems, particularly in small and medium size companies, but such schemes do not make more work available. The economic reforms, actually led to job reduction in India and therefore, opposed by labour organisations.

In developing economies, at the beginning of this century, people spent 35 per cent of their life span in paid employment. Today, this figure lies between 12 and 13 per cent social scientists predict that in future paid employment in Europe will only take up about six per cent of people's lives; they will be spending more time in education; paid working hours will shorten, and life expectancy will be longer. So the basic issue to consider is: what opportunities can be developed in addition to paid employment – for generating income, organising personal care, or simply giving life more meaning? All the above considerations take us to the next step: Where to go?

Promote 'People-Led' Decentralised Development

Sustainable development can be ensured only when the rural areas of developing countries will reap the benefits of development. Major thrust in current thinking about rural development in developing countries is that planning must be decentralised and rural people drawn into this process, so that decisions are better informed and more closely tailored to rural people's real needs.

In recent decades, three concepts have increasingly influenced rural development interventions. These are 'people-centred' development, 'public participation', and 'people-led' development. These terms are similar to one another and as a result they are often confusing, but they are not the same.

'People-centred' development refers to a planning mode through which experts generally seek to make proposed interventions acceptable to the final recipients. It seldom really alters what is done, but attempts to make the way in which it is imposed more recipient-friendly.

'Public participation' is an approach often used by development agencies to win acceptance of decisions that have already been made. It, thus, gives the recipients the sense of having been consulted, when, in fact, they might merely have been co-opted to accept something imposed from the outside.

'People-led' is a bottom-up approach to development and is no guarantee of success, nor of greater sustainability. It gives local people responsibility and the opportunity to take greater control of their own futures.

The way forward, therefore, should be guided by this principle.

However, putting such a 'people-led' approach into action requires following two changes in attitude on the part of those in power; the administrators, policy makers, planners and the 'experts':

(i) 'First shift' is to recognise and accept that local people are knowledgeable about their environment and circumstances; they do possess expertise that is valuable; and are thus able to contribute meaningfully to finding solutions to the problems they face. Strategies seeking to improve the rural situation should begin by building on this existing knowledge and expertise and by improving upon traditional, well-tested methods and practices. Devices and ideas being developed by experts external to the rural areas, should be evaluated by the potential consumers at an early stage, to allow them to be adapted to end users' needs. This approach should ensure that strategies would be appropriate to these needs and that users would be convinced of the benefits and thus more willingly adopt them.

(ii) 'Second shift' is attitude required to put into effect a people-led approach is possibly a little more difficult to achieve; that is the acceptance, again by those in power, that the approaches to resource management and conservation pursued in the past have essentially failed to secure a sustainable future. From this follows logically the need, for example, to relax the controls over resources, such as forests and woodlands, so that they can be entrusted more to the stewardship of those whose livelihoods are most dependent on them. In addition, in many cases, this entails foregoing some of the potential revenues from the exploitation of these resources.

Environmentally Sound Technologies (ESTs) hold the Key for Future Economic Development

Agenda-21 emphasises the importance of Environmentally Sound Technologies (ESTs). They "protect the environment, are less polluting, use all resources in a more sustainable manner, recycle more of their wastes and products, and handle residual wastes in a more acceptable manner than technologies for which they are substitutes". Agenda-21 also called on governments and other players to develop new funding mechanisms to accelerate the transfer of ESTs from the 'haves', mainly in the industrialised countries, to the 'have-nots', chiefly in the developing world. And it underlined the point by urging business leaders to give environmental management the highest priority in the move towards sustainable industrial practices, calling on business and industry to "develop techniques and technologies that reduce harmful environmental impacts". Agenda-21 also stressed that "new and efficient technologies will be essential to increase the capabilities, in particular of developing countries, to achieve sustainable development, sustain the world's economy, protect the environment, and alleviate poverty and human suffering".

What are ESTs? They cover a wide range of process and product technologies. Some, such as solar energy photovoltaics, fibre-optic devices, electric or ultra fuel-efficient vehicles, and biotechnology-based solutions, are available now, but are not yet economical enough to be commercially applied on a wide scale. In time, these and other emerging technologies will transform both production and consumption patterns, drive much of industry towards sustainable production, and eventually move some industrial sectors to the goal of zero emissions.

But the crucial point is that, meanwhile, many other ESTs, both 'end-of-pipe' and 'prevention technologies', are already available to achieve significant, sometimes dramatic, environmental improvements. Cleaner production technologies focus on achieving process and product change both to reduce pollution and also to prevent it occurring in the first place. End-of-pipe ESTs are more limited in scope, but they do control and reduce waste and emissions, as well as clean up pollution after it has happened. There is now a paradigm shift-taking place towards the cleaner production approach. As Agenda-21 makes clear, ESTs "are not just individual technologies, but total systems which include know-how, procedures, goods and services, and equipment as well as organisational and managerial procedures".

Industry, worldwide, though mainly in the industrialised countries, is using ESTs on a wider scale. One measure — and it is only one measure — is that the global market for environmental products such as water and wastewater treatment, waste management and recycling, air pollution control, noise reduction systems and services, is growing fast. It is increasing by around 5 per cent every year and is set to reach, conservatively, an annual total of US $ 300 billion by the year 2000. The United Nations Development Programme (UNDP) projects the global compliance market (including ESTs required to meet mandatory environmental standards) to reach US $ 500 billion a year by 2000. It forecasts that the future global market for environmental investments will expand to US $ 300-600 billion annually for pollution control, a total of US $ 1 trillion for power from 1993 to 2000, and a total of US $ 250 billion for energy conservation over the next 20 years.

What is difficult to calculate from these figures is the split between end-of-pipe and cleaner production technologies, although at the moment the market for the later is certainly much smaller. While end-of-pipe ESTs do help to reduce emissions, they are transitional rather than higher-term technology and should be used in conjunction with cleaner production approaches. Eventually, cleaner production process should take over and completely replace end-of-pipe technologies. Nonetheless, the market growth in ESTs does show that industry is moving to tackle environmental pollution problems.

There are several pressing requirements if further progress is to be achieved:

- to develop and commercialize new, advanced ESTs.
- to continue the shift from end-of-pipe approaches to cleaner production and eco-efficiency — that is, to move faster from pollution control to pollution prevention.
- to accelerate the knowledge and adoption of both new and existing ESTs throughout industry, in the industrialised world and, in particular, in the developing countries.
- to make the policy changes that will support the wider use of ESTs at the national and local level.
- to support the development of local ESTs.

UNEP GEO-1 Report states: "Appropriate technological improvements, which result in more effective use of natural resources, less waste, and fewer pollutant by-products, are required in all economic sectors. Truly global availability and worldwide application of best available and appropriate technology and production processes need to be ensured through the exchange and dissemination of know-how, skills and technology, and through appropriate finance mechanisms."

UNEP report highlighted three main barriers to accelerating the use of ESTs:

(i) lack of knowledge about the existence of commercially available technologies and the benefit they bring;

(ii) lack of financial resources to get them where they are needed;

(iii) lack of political action to create the right framework conditions for industry. While some governments are now using a mix of command-and-control policies and market-based incentives to push industry faster towards cleaner, more resource-efficient production, actual delivery on the political commitments made in Rio has, generally, fallen far short of what was promised.

The first two barriers apply particularly to developing countries though they are also a concern to small and medium-sized enterprises (SMEs) in many industrialised economies. The third relates to both developing and industrialised countries because there is a direct cause and effect between government policies and industry actions to improve environmental performance. There is no doubting the crucial role that government has in encouraging a faster take-up of ESTs as part of its move to promote sustainable development.

Conclusion

The above mentioned issues identify the course for sustainable development. There is no ideal single solution for different geographic locations. The basic question right now is 'as to how fast can we convince the governments and the people to change from current living practices to environment friendly practices'. It is possible to make fast progress by tackling today's environmental problems and following the path of sustainable development.

REFERENCES

Chaturvedi, Pradeep *et al*, *Business in a Cleaner Environment — The Indian Scenario*, World Environment Foundation, 1999.

Chaturvedi, Pradeep, World Congress on Sustainable Development, World Federation of Engineering Organisations, January 2000, Calcutta.

Strong, Maurice *et al*, *Food in the 21st Century: From Science to Sustainable Development*, CGIAR System Review Sect 9 (World Bank), 1999.

UNEP Report, *Global Environment Outlook*-1, 1997.

UNDP Report on Energy, *After Rio: Prospects and Challenges*, 1997.

UNEP Report, *Sustainable Business — Economic Development and Environmentally Sound Technologies*, 1998.

Visions for the 21st Century, Deutschland, December, 1999.

20

Sustainable Development and Governance: Future Directions

Introduction

Sustainable Development has been defined in a number of ways over the last decade. All actions as a follow-up of the United Nations Conference on Environment and Development, held in Rio De Janeiro in 1992, have created wider awareness of environment friendly solutions and the need for capacity building in each country. The Rio outcome as laid down in Agenda-21 provided the fundamental principles and the programme of action for achieving sustainable development. The United Nations Millennium Declaration and the outcome of the major UN Conferences and International Agreements since 1992 have evolved internationally agreed development goals; and have emphasised on expediting the realisation of the remaining goals of Agenda-21.

Components of Sustainable Development

These efforts will promote the integration of the *three mutually reinforcing pillars of sustainable development*:

(i) Economic growth,
(ii) Social development, and
(iii) Environmental protection.

Overarching Objectives of Sustainable Development

Poverty eradication and changing unsustainable patterns of

Paper presented at the 4th World Environment Congress organised by World Environment Foundation during June 6-8, 2002 at Palampur.

production and consumption are overarching objectives of sustainable development and an essential requirement for promoting environmental protection. Thus, sustainability on planet Earth can be achieved only when we tackle the issues of poverty and follow the sustainable patterns of production and consumption. This very concept forms the essence for discussion at the World Summit for Social Development, at Johannesburg, in August-September this year.

Benefits not Provided Equitably — The Major Gap of Implementation of Agenda-21: Need to Evolve Partnerships

We have also to recognize that the major gap in fulfilment of commitments for implementation of Agenda-21 has been that the benefits could not be provided equitably to all persons. It is now necessary that the World Summit on Sustainable Development should evolve the implementation strategy in such a fashion that the outcomes benefit all, particularly women, youth and vulnerable groups. The implementation should involve all relevant actors through partnerships, especially between governments of the North and the South on one hand, and between governments and major groups on the other, to achieve the widely shared goals of sustainable development. Such partnerships are key to pursuing sustainable development in a globalised world. So much for the sustainable development.

The Governance

The other part of this paper is the emphasis on governance. The Committee on Constitutional Reforms in India under Justice, M.N. Venkatachaliah, the former Chief Justice of India has defined governance to include governments, corporates and the civil society. Any discussion on good governance will have to focus on functioning and actions of these three groups distinctly and not of anyone. Good governance within each country and at the international level is essential for sustainable development. Corruption has been considered as the legacy of developing world and is reflected to be a sign of poor governance. Very little was known about the higher level corruption practiced in developed countries under the garb of efficient operational mechanism of large corporates. However, the ENRON debacle has brought forth that corporates all over the world have been evolving financial norms and procedures that did not reveal the correct picture to its shareholders and stakeholders. Maybe by the time the World Summit on Social Development will take place, the issue of corruption will be linked to the way the corporates in developed countries have embezzled large sums of common man's money. This is not to say that just because corruption takes place in the North, it can be tolerated or pursued in the South. Good governance which included good corporate governance and effectiveness of civil society actions — will remain the most important element for sustainable development.

At the domestic level, sound environmental, social and economic policies, democratic institutions responsive to the needs of the people, rule of law, anti-corruption measures, gender equity and enabling environment for investment are the basis for sustainable development. As a result of globalisation, external factors have become critical in determining the success or failure of developing countries in their national efforts. The gap between developed and developing countries point to the continued need for dynamic and enabling international economic environment supportive of international cooperation, particularly in the field of finance, technology transfer, debt and trade, full and effective participation of developing countries in global decision making, if the momentum for global progress towards sustainable development is to be maintained and increase.

This gives the perspective in which issues relating to sustainable development are to be seen. This will be dealt in detail later in the paper.

Sustainable Solutions for Mountain Eco Systems

This year has also been declared as the International Year of the Mountains, and the International Year of Eco-Tourism and UNESCO's International Year of Cultural Heritage. It is therefore relevant to discuss about these three aspects before getting into the major issues relating to sustainable development and governance. Mountain Eco System, include significant watershed resources, biological diversity, unique flora and fauna—and many are particularly fragile and vulnerable to the adverse effects of climate change. It is important to develop and promote programmes, policies and approaches that integrate environment, and economic and social components of sustainable mountain development and strengthen international cooperation particularly at the regional and sub-regional levels to agreements, treaties and conventions.

Some of *the important actions relating to sustainable solutions for mountain eco-system* include the following:

(i) Implement programmes to address, where appropriate, deforestation, erosion, land degradation, loss of bio-diversity, disruption of water flows and retreat of glaciers;
(ii) Develop and implement, where appropriate, policies and programmes, including public and private investments, that help eliminate inequities facing mountain communities, particularly for women;
(iii) Implement programmes to promote diversification and traditional mountain economies, sustainable livelihoods and small scale production systems, including better access to national and international markets, communications and transport planning, taking into account the particular sensitivity of mountains;
(iv) Promote full participation and involvement of mountain communities in decisions that affect them and to integrate indigenous knowledge, heritage and values in all development initiatives;
(v) Mobilize national and international resources with a view to promote investments, applied research and capacity building.

Promotion of sustainable tourism development includes non-consumptive and eco-tourism, taking into account the spirit of the International Year of the Eco-Tourism and the International Year of Cultural Heritage in order to increase the benefits from tourism resources for the population in host communities while maintaining the cultural and environmental integrity of the host communities and enhancing the protection of ecologically sensitive areas and natural heritage. The effort should be to promote sustainable tourism development and capacity building in order to contribute to the strengthening of rural and local communities. This would include following actions:

(i) Enhance international cooperation, foreign direct investment and partnerships with both private and public sectors at all levels;
(ii) Develop programmes that encourage people to participate

in eco-tourism and enhance stakeholder co-operation in tourism development and heritage preservation to improve the protection of the environment, natural resources and cultural heritage;

(iii) Provide technical assistance to developing countries and countries with economies in transition to support sustainable tourism business development and investment, tourism awareness programmes to improve domestic tourism, and to stimulate entrepreneurial development;

(iv) Assist host communities to manage visitation to their tourism attractions for their maximum financial benefit whilst ensuring the least negative impact on and risks for their traditions, culture and living environment. It is further recommended that the World Tourism Organisation and other relevant organisations facilitate the implementation in their Member States;

(v) Promote the diversification of the economic activities, including through the facilitation of access to markets and commercial information, and participation of emerging local enterprises, especially SMEs.

Overarching Objectives of Sustainable Development

Poverty eradication and changing unsustainable pattern of production and consumption were introduced as overarching objectives of sustainable development and an essential requirement for promoting environmental protection. It is, therefore, necessary that these two issues are dealt more elaborately in the paper.

Poverty Eradication

Eradicating poverty is the greatest global challenge facing the world today and an indispensable requirement for sustainable development, particularly for developing countries. While each country has primary responsibility for its own sustainable development and poverty eradication, concerted and concrete measures are required to achieve the internationally agreed poverty related targets. Actions at the international, regional and national levels are required to accomplish these targets. The Millennium Development Goal related to poverty aims at reducing by half, by 2015, of the proportioned people whose income is below US $ 1 PPP per day, and the number of people suffering from hunger, and proportion of people without access to safe drinking water.

Various international, regional and national actions will include the following:

(i) Establish a World Solidarity Fund for Poverty Eradication and the Promotion of Human Development in the poorest regions of the world, pursuant to modalities to be determined by the UN General Assembly.

(ii) Develop national programme for sustainable development and community development to promote the empowerment of people living in poverty and their organisations. These programmes should reflect their priorities, and enable them to increase access to productive resources, public services and institutions, in particular land, water, employment opportunities, credit, education and health.

(iii) Promote women's participation in decision making at all levels, mainstreaming gender perspectives in all policies and strategies.

(iv) Deliver basic health services for all and reduce environmental health threats, taking into account the linkages between poverty, health and environment, with provision of financial resources, technical assistance and knowledge transfer to developing countries and countries with economies in transition.

(v) Ensure children everywhere, boys and girls alike, will be able to complete a full course of primary schooling and they will have equal access to all levels of education.

(vi) Provide access to other agricultural resources, land tenure arrangement that recognize and protect indigenous and common property resource management systems.

(vii) Build basic rural infrastructure, diversify the economy, improve access to markets and to credit for the rural poor to support sustainable agriculture and rural development.

(viii) Transfer basic sustainable agricultural techniques and knowledge, including natural resource management, to small and medium scale farmers, fishermen and the rural poor, including through multi-stakeholder approaches and public—private partnerships aimed at increasing agricultural production and food security.

(ix) Increase food availability and affordability, including through harvest and food technology and management, as well as equitable and efficient distribution system by promoting community based partnerships linking urban and rural people enterprise.

Linked to it is the action programme to halve by 2015 the proportion of people lacking access to improved sanitation and modern energy services. Certain actions needed for the energy sector include the following:

(i) Utilise financial instruments and mechanisms to provide financial resources to developing countries to meet their capacity needs; and strengthen national institutions in energy, including promoting energy efficiency, advanced fossil fuel technology and renewable energy.

(ii) Improve access to energy services in rural and semi-urban areas through rural electrification and decentralised energy systems. The definition of the village electrification needs to be modified from one bulb in a village to at least 30 per cent electrified houses in a village.

(iii) Develop regional plans of action to facilitate cross-border energy trade, including the interconnection of electricity grids and oil and natural gas pipelines.

(iv) Improve access to modern biomass technologies and fuelwood sources and supplies, and commercialise biomass operations, including the use of agricultural residues, where such practices are sustainable.

(v) Strengthen, and where appropriate, establish policies on energy for rural development, including, as appropriate, regulatory systems to promote access to energy in rural and semi-urban areas.

The next aspect in poverty eradication is strengthening contribution of industrial development and sustainable natural resource management. Such actions will include the following:

(i) Provide assistance to enhance industrial productivity and competitiveness as well as industrial development in developing countries;

(ii) Promote the development of micro, small and medium size enterprises, with a special focus on agro-industry as a provider of livelihoods for rural communities;
(iii) Provide financial and technological support to rural communities of developing countries to enable them to benefit from small-scale mining ventures;
(iv) Provide support to developing countries for the development of low-cost technologies that conserve fuel for cooking and water heating; and
(v) Remove unnecessary regulatory and other obstacles for micro enterprises and the informal sector.

Changing Unsustainable Patterns for Consumption and Production

Fundamental changes in the way societies produce and consume are indispensable for achieving global sustainable development. All countries have to strive to promote sustainable consumption and production patterns, and with all countries benefiting from the process, based on the principle of common, but differentiated responsibilities. Governments, relevant international organisations, the private sector and all major groups should play a critical role in striving to change unsustainable consumption and production patterns. This will basically include the following:

(i) Developing a ten-year work programme for improving resource efficiency in order to promote social and economic development within the carrying capacities of eco systems, while reducing resource degradation;
(ii) Increase investments in cleaner production and eco-efficiency through incentives and support schemes;
(iii) Enhance corporate, environmental and social responsibility and accountability;
(iv) Provide training for relevant authorities at all levels to take sustainability considerations into account in decision-making, including on national investments in infrastructure, business development and public procurement;
(v) Promote the implementation of recommendations that relate to energy for sustainable development, taking into account that energy is central to achieving the goals of sustainable development;

(vi) Promote an integrated approach to policy making at national and regional levels for transport services and systems to promote sustainable development, including policies and planning for land use, infrastructure, public transport systems and goods delivery networks, with a view to providing efficient transportation, reducing energy consumption and pollution, reducing congestion, limiting urban sprawl, and promoting long term sustainable development taking into account national priorities and circumstances; and

(vii) Prevent and minimise, waste and to maximise reuse and recycling, with participation of government and all stakeholders, in order to improve resource efficiency.

Governance for Sustainable Development

The potential of globalisation to promote *sustainable development for all* still remains a distant dream. The Secretary General of the United Nations in his Millennium Report had warned that unless the globalisation process brings the poor centre stage it will lose its meaning and effectiveness. Good governance practices are to evolve to ensure national economic and social policies in such a fashion so that sustainable development becomes a feasible option. Governments need to announce policies that lead to wealth creation that in turn supports social development. Some of the actions required include:

(i) Adoption and implementation of coherent and sound micro-economic polices and development of institutional capacities;

(ii) Create and open, equitable, rule based, predictable and non-discriminatory multilateral trading and financial system that benefits all countries in the pursuit of sustainable development;

(iii) Make foreign direct investment more supportive of sustainable development and encourage international and regional institutions as well as institutions in source countries to promote increasing investment flows to developing countries and to assist developing countries in their efforts to create a conducive, domestic and enabling environment;

(iv) Promote corporate responsibility and accountability and the exchange of good practices including through public-private partnerships and voluntary initiatives building, *inter alia*, on the UN Global Compact and the OECD Guidelines for Multilateral Enterprise as appropriate, as well as by using tools such as Environmental Management Accounting and Environmental Reporting; and

(v) Enhance Corporate responsibility and accountability and the exchange of good practices, including through public/private partnerships and voluntary initiatives.

Enhancing corporate environmental and social responsibility and accountability will include the following elements:

(a) Encourage industry to improve social and environmental performance through voluntary initiatives, including environmental management systems, codes of conduct, certification, and public reporting on environmental and social issues, taking into account such initiatives as the International Organisation for Standardisation (IOS) standards and the Global Reporting Initiative (GRI) guidelines on sustainability reporting, bearing in mind Principle 11 of the Rio Declaration;

(b) Encourage dialogue between enterprises and the communities in which they operate and other stakeholders;

(c) Encourage financial institutions to incorporate sustainability considerations into their decision-making processes; and

(d) Develop workplace-based partnerships and programmes, including training and education programmes.

Conclusion

Sustainable development is probably the most daunting challenge that humanity has ever faced, and achieving it requires that the fundamental issues be addressed immediately at local, regional and global levels. Bridging the development gap between the North and the South, and alleviating poverty to provide a more equitable and sustainable future for all, requires novel integrated approaches that fully incorporate existing and new scientific knowledge. Whatever the cultural, geographical, socio-economic and environmental setting, a strong partnership between the S&T community and other members of civil society, the private sector (or the corporates) and government is a fundamental prerequisite to sustainable development.

21

Future of Environmental Governance

Introduction

Environmental and social development of the Himalayan communities is a significant move as the fragile ecosystem of the Himalayas has been very often referred to be permanently destroyed by development process in the past. The Third World Congress on Environment Management has most appropriately focused on the issues that will lead to development, economic prosperity and environment preservation of Himalayan region.

Very little has been thought in the past of how bio-resources of Himalayas can also be exploited and re-generated to bring prosperity in this region without disturbing the ecological balance. WCEM being held in the penultimate year of the Rio+10 Conference, it is but natural to link to the development process that has taken place over the last 30 years.

As environmental issues had been politicised and the environmentalist groups have tried to secure support, there has been a tendency to stress the sensational. Environmental issues have been projected in terms of crisis and catastrophe. The politicisation of environment has brought the greenhouse effect as common man's talk and made the ozone hole a headline in daily newspapers. Anti-dam lobby is projecting greenhouse gas emissions even from the surface of water held in large dams.

This may not be wrong in itself but let us not forget that the whole world cannot be left behind these global issues and the tragedy of increasing degradation of our land and water resources, the growing scarcity of drinking water and lack of fuelwood will affect the well

---

Paper presented at the Third World Environment Congress organised by World Environment Foundation during June 3-5, 2001 at Palampur.

being of a large mass for common people. The greenhouse effect may be a few decades away, but there is not enough fuelwood even now and its lack is leading to denudation of our forests. We do not need to sensationalise environment. Environmental deterioration is a matter of people's daily experience; once they see that connection, they will take environmental issues seriously. But then, they will ask for solution and that is where work begins. We have to decide whether we respond to these needs or else let some other groups of activists with their own agenda project development aspects as anti-environment. Facts can be processed and re-processed and presented to suit any theory positive or negative depending on the agenda one wishes to pursue. Those who look at positive linkage between environment and development, therefore, have the responsibility to project strongly about the positive role that development has in relation to environmental preservation, otherwise the negative impact will catch attention.

The environmental issues are intricate and interwoven. Wind currents, rain pattern and river streams carry pollutants hundreds or even thousands of kilometres from their sources. Some 200 river basins are shared by two or more countries. On an even larger scale, the global environmental problems of ozone depletion, climate change, deforestation, and the loss of the earth's biological diversity threaten all nations. Anyone country acting alone is powerless in the face of these problems.

Need for Environmental Governance

National sovereignty—the power of a country to control events within its territory—has not much of its meaning in today's world, where borders are routinely breached by pollution, international trade, financial flows, and refugees. Increasingly, they may be eroded by such forces as climatic warming, migrations, and the depletion of the earth's ozone shield. Because all these forces can affect environmental trends, international treaties and institutions are proving ever more critical to addressing ecological threats. Nations are in effect ceding portions of their sovereignty to the international community, and beginning to create a new system of international environmental governance as a means of solving otherwise-unmanageable problems. Certain developing countries oppose such an international arrangement and do not take benefit of international cooperation.

International efforts to protect the natural environment date at least to the 1870s, when Switzerland tried to establish a regional agreement to protect the nesting sites of migratory birds. But the move to internationalize environmental policymaking did not gain serious momentum until the 1970s. Up to the time of UN Conference on Environment Development in 1992, at Rio, governments had adopted more than 170 environmental treaties on subjects of shared concern; acid rain, ocean pollution, endangered species protection, hazardous wastes export, and the preservation of Antarctica, as well as ozone depletion. More than two-third of these agreements had been reached since the landmark U.N. Conference on the Human Environment was held in Stockholm in 1972. Governments have also deployed international institutions such as the United Nations Environment Programme (UNEP), created at the Stockholm conference, to negotiate treaties and help make them work. Let's wait and see as to how many more treaties come in effect by the time the UN World Summit on Sustainable Development takes place in June 2002 at Johannesburg.

Ecological issues have in the space of a few years become a central feature of international relations. In the late eighties, when global warming first emerged as a pressing international priority, Gro Harlem Bruntland, Mikhail Gorbachev, Margaret Thatcher, and other world leaders seemed at times to be competing to be the greenest actors on the world state. At the 1989 G-7 summit meeting, environmental questions dominated the discussions and the communiqué.

However, rhetoric about the need for international environmental governance has far outweighed effective action. In the year preceding the June, 1992 United Nations Conference on Environment and Development (UNCED), it became clear that this so-called "Earth Summit" would be a critical test of resolve for world leaders. The agenda for the Rio meeting promised historic breakthrough: treaties covering global warming and the loss of biodiversity; an action plan covering oceans, forests, toxic chemicals, and other issues; and a "Earth Charter" to lay down a set of basic international environmental principles. But it was also clear that such far-reaching agreements and resolutions would not yield progress unless the participating governments could find adequate ways to implement and pay for them.

The current system of environmental treaties fall short on several counts. Existing treaties and agreements are often weak and unenforceable. The following two major conventions have not been very effective in implementation:

(i) UN Framework Convention on Biological Resources; and
(ii) UN Framework Convention on Climate Change.

And governments have not yet ceded to international organisations either the powers or the financial resources they will need to stave off ecological disaster. Most of today's international institutions were fashioned in the forties and fifties, when environmental issues were not yet even on national agenda. Fundamental reforms will be needed if yesterday's institutions are to respond to today's – and tomorrow's — challenges.

The United Nations Framework Convention on Climate Change UNFCC, has served as a major treaty to create awareness and have a uniformly acceptable approach on a global basis. Not all the countries signed UNFCC so far. Subsequently, the implementation strategy for making environment clean was worked out within the framework of the Kyoto Protocol. The provisions of the Kyoto Protocol generated heat and debate and were not endorsed by the member countries in sufficient numbers to ratify it. The Joint Implementation, JI, project did start but the Clean Development Mechanism, CDM could not even be formulated and initiated.

The global efforts to create environment fund, popularly known as Global Environment Fund, GEF, was a massive exercise. Whereas World Bank, UNDP and UNEP, each played its role, three parallel channels for funding exists under these three organisations. However, the procedures for accessing such funds are so time consuming and difficult that their impact in short-term is really not visible. The basic concept of GEF funding was to promote participation of private industries and infrastructures to support environmental cleanliness through involvement of society.

The procedures followed during the last 9 years after the Rio Conference have neither satisfied the government agencies, nor the non-government organisations. During these decades the NGO movement has become much stronger and they are now voicing their concern even at the policy level. Realising the strength of the NGO movement, the United Nations has now made it mandatory for all its agencies to involve NGOs in all their development conferences and activities. This open participation of NGOs has led to a new approach on the international arena, wherein people's opinions are also been directly reflected in the decision making process.

In addition to protecting the prerogatives of local or regional governments, wherever possible, international environmental governance will need to protect the interests of developing world, in which perceived priorities are different from those of the dominant industrial nations. During the months preceding the UNCED, there were disturbing signs that these divergent perceptions were creating new antagonisms. The differences became sharper after the Rio Conference. To resolve these, the international environmental agenda needs to be linked with the broader economic and development context of which it is a part. Encouraging full developing-country participation in international environmental governance will require more debt relief, reduction of trade barriers, and short-term aid for the hungry or poor. Without these measures, developing countries may be unable to focus on meeting their longer-term environmental commitments. Such issues have still not been tackled.

International institutional development is not keeping pace with either the world's ever-growing interdependence or the rapidly deteriorating conditions of the Earth. Existing treaties and declarations represent embryonic attempts to set the world on a sustainable course, and indeed they have led to some important gains. But in the long run, to achieve only modest success in this realm will not be enough. Real success in reversing environmental decline will not be possible until more fundamental changes are brought in the mechanisms of governance themselves.

Strengthening Environmental Institutions

To meet the current highly laborious and piece-meal process of international environmental governance, stronger institutions are needed to help negotiate treaties, implement them, settle disputes over their terms, and dispense the financial and technological assistance required to make them work.

The time may be ripe not only to adopt majority-voting provisions within the individual governing bodies for treaties, but to establish a central negotiating forum. Under existing practice, a new international committee—such as the International Negotiating Committee on Climate Change—is created nearly every time a major treaty is up for discussion. A unified forum could greatly improve the efficiency and dispatch with which rapidly-escalating global problems are addressed.

It is more relevant to talk of reforming economic institutions to fit into the changing economic and environmental regime.

Reforming Economic Institutions

In national governments, the finance and commerce ministries often hold sway over environmental protection ministries in interagency struggles. At the international level as well, power rests with the institutions charged with managing the world economy. Strengthening global environmental governance will thus depend in large measure on substantially reforming today's leading international economic institutions—the World Bank, the International Monetary Fund (IMF), and the World Trade Orgnisation (earlier known as) General Agreement on Tariffs and Trade (GATT).

These organisations have achieved considerable success in the task granted them by governments when they were created after World War-II—stabilising the world monetary system and preventing a return to the protectionism of the thirties that contributed to the global depression. But the environment was not recognised to be inextricably linked with the economy at the time, as it is beginning to be now. The globalisation of the world economy, combined with the power of the World Bank, IMF, and WTO to affect policies, means that a growing amount of *de facto* environmental governance is now taking place through the "back door"—*via* these organisations. As these approaches are based on systems and approaches operational in developed countries, many a developing countries are not enthused.

Environment and Trade

There are number of ways in which free trade and environmental goals conflict. Product standards have the effect of discriminating against foreign made products — such as cars not meeting domestic automobile fuel efficiency standards, or beverages sold in non-returnable bottles — whether or not this was there intent. Other important conservation strategies also can constrain trade.

Import taxes on products made in countries with lax environmental standards have also been attacked as trade barriers. The usual motivation for these taxes is to ensure that domestic producers are not put at a competitive disadvantage by having to meet strict environmental standards. But these levies also help prevent the export of hazardous industries to countries where regulation is lax. Such countries are in essence subsidising domestic industries at the environment's expense, a practice that is coming to be known as ecological dumping.

The most logical solution to the trade and environment dilemma is to negotiate strict international environmental standards that require companies in every country to play by the same rules. This level playing field would ensure both free trade and environmental protection in an increasing global economy. But there are practical difficulties in negotiating a common policy. Forging a consensus among countries that have varying commitments to environmental protection and are at different stages of economic development is difficult.

Environmental management systems, popularly known as ISO 14000 standards have greatly created awareness in industrialised or industrialising countries and those seeking a rightful place in the globalised economy. The situation is not uniform nor is the countrywide application uniform or logical. Realising the need for building quality and associated elements of environment management the new ISO 19000 standards will help. These standards are applicable to biological sector, industry sector as well as the service sector. Conformance to ISO standards on environment management systems provides an opportunity for industries in different parts of the world to access new markets. By the time, the Rio+10 Conference will take place in Johannesburg, the environment management systems would have created a level playing field for players from different parts of the world.

Relevant UN Expert Meetings

Two expert meetings took place in preparation for RIO+10 and are relevant for our discussion. These are as follows:

Expert Meeting on Cleaner Technologies

(i) The second regional expert meeting on National Cleaner Technology Strategies was held in Bratislava, Slovakia, in June 2000 for the countries of Eastern Europe and the former Soviet Union, organised by the Division for Sustainable Development and the Slovak National Cleaner Production Centre and hosted by the Government of Slovakia. The meeting focused on ways to promote investment in production technologies that increase productivity, conserve natural resources, and reduce environmental impacts.

It was agreed that many profitable investments in new, cleaner technologies are not being made because enterprises and financial institutions do not recognize to cost reductions and revenue increases resulting from greater efficiency of resource use, reduced waste and pollution, and increased sales in environment-sensitive markets. Enterprises and financial institutions need training in the use of investment appraisal and project formulation that identify the financial benefits to enterprises of cleaner production. Cleaner production is a particular priority in Eastern Europe because of the importance of the environmentally-sensitive European Union market.

In addition to conventional sources of investment finance, there are a number of sources of investment finance specially oriented towards environmentally sound production, such as the European Bank for Reconstruction and Development and the Global Environment Facility. These sources offer favourable conditions for cleaner technology investments and are looking for well-formulated project proposals. These sources, however, often have specific target countries and criteria, and government agencies and enterprises need assistance in identifying the best sources in formulating proposals meeting in criteria of those sources.

The first regional expert meeting was held in Bogota, Colombia, in October, 1999, for the countries of Latin America and the Caribbean. Regional meetings are being planned for Africa, Asia and the Pacific.

Expert Meeting on Environmental Accounting

(ii) The third meeting of an expert group on Environmental Management Accounting (EMA) was held in Bonn, Germany, in November 2000, organised by the Division for Sustainable Development and hosted by the German Federal Ministry of the Environment and the Federal Ministry of Education and Research. The meeting focused on ways in which governments could promote the use of EMA by enterprises in order to improve both the economic and environmental performance of industry. There is a growing recognition that conventional corporate accounting systems do not adequately identify the costs of waste and pollution. Environmental costs, such as waste management and legal and regulatory expenses, are generally allocated to overhead and then divided among all production lines, with the result that "dirty" production lines are in effect subsidised by cleaner lines. Environmental management accounting systems that identify environmental costs by product line can help in identifying cost-effective ways to reduce such costs, thus increasing profits while reducing environmental impacts.

The meeting reviewed draft studies on EMA systems, linkages between corporate EMA and other environmental management systems, and government policies for promoting corporate EMA. Consultants preparing these reports have developed an approach to EMA combining improved cost accounting procedures and material flow analysis in order to identify the savings that can be realised through cleaner, more efficient production methods. The studies are to be published and distributed in early 2001.

Focus of Commission on Sustainable Development

The Commission on Sustainable Development, CSD, has focused on following issues in its 5-year Work Programme (1998-2002):

Year	*Themes*
1998	Freshwater Management Transfer of Technology/Capacity-building/ Education/Science Industry Sustainable Development of SIDS
1999	Oceans and Seas Consumption and Production Patterns Tourism

Comprehensive review of the Programme of Action for the Sustainable Development of SIDS (September 1999)

2000	Land Resources Financial Resources/Trade and Investment/ Economic Growth Sustainable Agriculture Day of Indigenous People
2001	Atmosphere; Energy Information for Decision-making and Participation International Cooperation for an Enabling Environment Energy; Transport
2002	Comprehensive review of Agenda-21 implementation (Rio+10)

Conclusion

The situation right now is that after the CSD 9 meeting in April this year, in New York, the policy makers realised the need for environment preservation through involvement of business interests and the NGOs. Globally that will only be the approach that will help in creating larger opportunities for environment improvement whereas expanding the scope of the business opportunities. Environment and business are now considered complementary to one another and the environmental concerns are no more a barrier to trade. The emerging standards for process and products help to make environment friendly business and trade a reality. The same approach is to be implemented to make sure that the development of the Himalayan region and the Himalayan communities takes place in environment-friendly fashion.

22

Globalisation, Environment Management and Quality of Life

Introduction

Globalisation of Indian economy has been the catchword of the last decade of the millennium. Proponents of the concepts who strongly believe that India can earn its position in the global economy and earn the foreign exchange necessary for building India's socio-economic structure, had lot to talk about and convince the policy makers and the general public. An equally strong opposition has been critical of following the path of globalisation as that aims at removing subsidies, correcting prices and shifting the balance in favour of large and modem companies. The critics also firmly believe that an approach to globalisation results in loss of employment opportunities, and weakening the local manufacturing and R&D base, thereby demolishing the self-reliance of the country. Liberalisation and globalisation as a process of change do not in themselves guarantee growth and economic development at desired levels.

Liberalisation has given rise to a new culture which follows a course different from planned economy. The critical difference between planning and liberalisation is: planning assumes that outcomes are predictable, while liberalisation facilitates the unpredictable. It does so by unshackling the energy of innovation of millions of people unknown to planners, of letting them play around with ideas unthought of. Such unknown people and ideas have transformed other countries and can also transform India. It enables unknown to do the unthought of and become unstoppable.

The whole approach to globalisation of Indian economy and its

---

Paper presented at the Second World Congress on Environment Management organised by World Environment Foundation during June 2-3, 2000 at New Delhi.

impact on Indian efforts at innovation, environment management and poverty alleviation need a total dispassionate study.

What is Globalisation?

Mr. Kofi Annan, Secretary General, United Nations, in his Millennium report mentions:

"Globalisation is transforming the world as we enter the 21st Century. In this new era, peoples' actions constantly — if often unwittingly — affect the lives of others living far away. Globalisation offers great opportunity, but at present its benefits are very unevenly distributed, while its costs are borne by all.

Thus the central challenge we face today is to ensure that globalisation becomes a positive force for all the world's people, instead of leaving billions behind in squalor. "Inclusive globalisation" must be built on the great enabling force of the market, but market forces alone will not achieve it. It requires a broader effort to create a shared future, based upon our common humanity in all its diversity.

That, in turn, requires that we think afresh about how we manage our joint activities and our shared interests, for many challenges that we confront today are beyond the reach of any state to meet on its own. At the national level, we must govern better, and at the international level we must learn to govern better together. Effective states are essential for both tasks, and their capacity for both needs strengthening. We must also adapt international institutions, through which states govern together, to the realities of the new era. We must form coalitions for change, often with partners, well beyond the precincts of official dump.

No shift in the way we think or act can be more critical than this: we must put people at the centre of everything we do. No calling is more noble, and no responsibility greater, than that of enabling men, women and children, in cities and villages around the world, to make their lives better. Only when that begins to happen will we know that globalisation is indeed becoming inclusive allowing everyone to share its opportunities."

Post-1945 Economy

Globalisation is not new, but the present era has distinctive features. Shrinking space, shrinking time and disappearing borders are linking people's lives more deeply, more intensely, more immediately than ever before. The phenomena of globalisation can be traced to over thousand years when countries started trading with each other. The earlier concept of globalisation was restricted to three basic elements: physical capital in terms of plant and machinery; labour and technology. The concept continued well into 1960s when theories in economics got prominence, when new interpretation of globalisation of economy emerged. 1960s saw the emergence of information technology and its integration with capital market. This combination brought forth a new process of globalisation, and the movement is often termed as financial capital. Thus, globalisation starts with the concept of 'financial capital' in terms of capital invested in the capital market — mainly foreign direct investments in emerging markets.

The post-1945 era realised the need for economic inter-dependence. The leaders chose a course of openness and cooperation. The United Nations, the Bretton Woods institutions and the General Agreement on Tariffs and Trade (later incorporated into the World Trade Organisation) and a host of other organisations were established, whose job was to make the overall systems work. In the industrialized countries, domestic support for open market was secured by constructing social safety nets and providing adjustment assistance to adversely affected groups and industries. The situation as it existed immediately after the Second World War was well served by the older concept of globalisation.

The Secretary General, UN commented on status of globalisation as: "The crux of today's problem, where globalisation has to support build up of economies of a large number of independent developing countries is that while the post-war multilateral system made it possible for the globalisation to emerge and flourish, globalisation, in turn, has progressively rendered its designs antiquated. Simply put, our post-war institutions were built for an international world, but we now live in a global world. Responding effectively to this shift is the core institutional challenge for world leaders today.

The post-war institutional arrangements were premised on a world made up of separate national economies, engaged in external transactions, conducted at arms length. Globalisation contradicts each of these expectations. It is hardly surprising, therefore that the trade regime is under such stress — it increasingly deals with traditionally "domestic" matters rather than border barriers. Thus, calls for a new financial architecture are so insistent.

Characteristics of Present Globalisation

The Human Development Report 1999 of UNDP defines characteristics of modern day globalisation as follows:

(i) *New markets* — Foreign exchange and capital markets linked globally, operating 24 hours a day, with dealings at a distance in real time.
(ii) *New tools* — Internet links, cellular phones, media network.
(iii) *New actors* — WTO with authority over national governments, the multinational corporations with more economic power than many states, the global networks of non-governmental organisations and other groups that transcend national boundaries.
(iv) *New rules* — Multilateral agreements on trade, services and intellectual properties backed by strong and enforcement mechanisms and more binding for national governments, reducing the scope for national policy.

Does Globalisation offer Opportunities to Millions of Poor?

Thus, the era of globalisation is opening many opportunities for millions of people around the world. Increased trade, new technologies, foreign investments, expanding media and internet connections are fueling economic growth and human advance. More than $ 1.5 trillion is now exchanged in the world's currency markets each day and nearly a fifth of the goods and services each year are traded. Globalisation is far beyond just the flow of money and commodities. It is growing interdependence of the world's people. Globalisation is a process integrating not just the economy, but culture, technology and governance. People everywhere are connected — affected by events in far corners of the world. Global markets, global ideas and global solidarity can enrich the lives of people everywhere, greatly expanding their choices. The growing interdependence of people's lives calls for shared values and the shared commitment to the human development of all.

Globalisation can become an effective tool to offer growth opportunity to millions of poor. But, not by excluding them from the markets. This will be feasible only through inclusive globalisation, where the poor have a role to play.

The enormous potential of globalisation needs to be tuned to eradicate poverty in the 21st Century — to continue the unprecedented progress in the 21st Century. UNDP identifies that to achieve this the challenge of globalisation in the new century is to find the rules and institutions for stronger governance to preserve the advances of global markets and competitions, but also to provide enough space for human, community and environmental resources to ensure that globalisation works for people — not just for profit. Globalisation has to evolve with:

(i) *Ethics* — less violation of human rights, not more.
(ii) *Equity* — less disparity within and between nations, not more.
(iii) *Inclusion* — less marginalisation of people and countries, not more.
(iv) *Human Security* — less instability of societies and less vulnerability of people, not more.
(v) *Sustainability* — less environmental destruction not more.
(vi) *Development* — less poverty and deprivation, not more.

The opportunities and benefits of globalisation need to be shared much more widely. If global opportunities are not shared better, the failed growth of the last decades will continue. More than 80 countries still have per capita incomes lower than they were a decade ago; while 40 countries have sustained average per capita income growth of more than 3 per cent a year since 1990, 55 countries, mostly in Sub-Saharan Africa and East Europe and the Commonwealth of Independent States, have had declining per capita incomes. Many people are also missing on employment opportunities. Inequality between countries has also increased. The income gap between the one-fifth of the world's people living in the richest countries; and the one-fifth in the poorest was 74 to 1 in 1997 up from 60 to 1 in 1990, and 30 to 1 in 1960. These are the signs that often carry the governments in developing countries to think away from globalisation.

Globalisation has also created new threats, mainly in the following categories:

(i) Financial volatility and economic insecurity.
(ii) Job and income insecurity.
(iii) Health insecurity.

(iv) Cultural insecurity.
(v) Personal insecurity.
(vi) Environmental insecurity.
(vii) Political and community insecurity.

Thus, the public perception is that globalisation has deleterious effect in developing countries. This has to be countered by evolving new concept of globalisation.

Observations on Recent Growth Patterns in India (in the Era of Liberalisation)

The general impression is that the rural-urban divide is widening. The observation of the National Sample Survey on household consumer expenditure is an eye opene.·. There has been a significant decline of 6.7 per cent in rural spending and an accompanying 4.6 per cent increase in urban spending during 1991-98. Rural poverty has risen by 3.42 per cent since the beginning of reforms. Urban poverty, however, has declined by just over 1 per cent.

One should not get the wrong signal that the economic reforms have fuelled rural poverty. The 54th round of the NSS (data collected during January-June 1998) estimates an average monthly per capita expenditure of Rs.455. In urban areas, the figure is Rs.684; and in the rural areas it is only Rs. 382. In villages, 61 per cent of the expenditure is estimated to be on food; in towns it is 50 per cent. "The poorer the household, the higher the expenditure on food. It does not reflect intake of improved quality of food but only that the poor is being adversely affected by the rising prices." The share of expense on food has actually fallen. The survey shows that non-food expenditure increased from 35 per cent, during 1987-92, to 39 per cent during 1996-98 in rural India, and from 43 to 49 per cent in urban centres.

The Planning Commission had indicated that the 80s recorded a fall in the poverty line by more than 10 per cent, the absolute numbers falling by 50 million. In the reform period, however, poverty has risen by 130 million, an increase of 9 per cent over 1989-90. Experts point out a paradox: Despite a 5.6 per cent growth in Gross Domestic Product (GDP) between 1990-91 and 1996-97, poverty increased by one per cent a year. In contrast, it fell by 3.1 per cent annually with a lower GDP growth between 1983 and 1990. In 1993-94, GDP grew by 6.9 per cent, yet poverty increased by 2 per cent. The inference is that reforms have reversed the process of poverty reduction.

But the irony is that poverty increased from 36 per cent in 1993 to 40 per cent in 1998. Yet, according to a survey by the National Council for Applied Economic Research (NCAER), urban income, which was 1.82 times higher than rural, in 1975-76, had become 2.1 times higher by 1994. Farm sector income declined from 37.9 to 20.5 per cent.

This observation only shows that the money allocated for rural development work has not been appropriately utilised. It is difficult to believe that the funds are not reaching the needy.

In certain observations (by a number of foreign aid agencies) it has been reflected that the aid allocations are not fully and effectively utilised because engineering inputs required for poverty alleviation programmes are not fully available in India. Local solutions for problems are desired. These agencies have interacted with leading R&D institutions — but of no avail. Such aid comes only from the foreign governments, or international agencies and makes it obligatory for donors to approach only the government controlled R&D institutions. Such agencies very often do not have appropriately trained manpower. As the local solutions are not found to have the potential of job creation out of such donations, the international agency allocations are not fully utilised. Whereas liberalisation has given higher GDP growth rate, it has not created necessary higher employment opportunities.

The outright and gross negligence of the poor in urban and rural areas has kept them out of the market place. They do not get benefit of the liberalised economy and also do not contribute to growth of globalised Indian economy. The poor, thus, remain market excluded. The government is reducing subsidies (and support) thereby pushing the poor into deeper poverty trap. If the government is not able to meet all funding requirements for uplift of the poor, then the poor need to be enabled to benefit from the market mechanism. An 'inclusive globalisation' is an effective approach for poverty eradication.

Need for Globalisation with a Human Face — The Inclusive Globalisation

'Inclusive globalisation' needs to tackle the issue of technology development for human development and eradication of poverty. The potential of the new technologies has to be tapped, which will call for public investments to fulfil the needs of the poor people and poor countries in everything from seeds to computers. An international programme can be considered to support this, based on the model of the Consultative Group on International Agricultural Research (CGIAR). Narrowing the gap between rich and poor; and narrowing the extremes between countries should become explicit goals of globalisation. This would complement the goals for poverty reduction and social advance agreed to in the global conferences of the last decade of the millennium.

Dr. C.K. Prahlad, the Management Guru, in his address to the CII Economic Summit, in New Delhi on January 10, 2000, has given a message to create a 'consumer market' out of the poor (instead of only focusing on niche market). He called on the Indian CEOs to think out of the box and make more money out of the poor. Dr. Prahland advised that the Indian industry have to use its 'problems' as 'advantages' to build a globally competitive business. The challenge today is to target bottom of the pyramid consumer.

He said, "you have to create a consumer market of the bottom layers of the Indian poor. If Indian entrepreneurs do not innovate, MNCs could capture the domestic markets. There is a strong possibility of turning water, food and primary education into a business model."

In simple words he advised industry leaders to create avenues for new wealth generation by building consumer markets, comprising the poor. At present the poor are market excluded, because they do not have the purchasing capacity to enter the market. The advice given by Dr. Prahlad is timely and indicates the urgent need to bring the poor into the 'market economy'. India cannot create 'islands of wealth' in the 'ocean of poverty' and still grow equitably.

Environmental Security

The world population is projected to increase to nearly 9 billion by 2050, from the current 6 billion, with the potential of irreparable environmental damage. One of two jobs worldwide — in agriculture, forestry and fisheries — depends directly on the sustainability of ecosystems. Even more important, so does the planet's health — and our own depend directly on the sustainability of ecosystems.

Environmental Sustainability is Everybody's Challenge

In the rich countries, the by-products of industrial and agribusiness production poison soils and waterways. *In the developing countries*, massive deforestation, harmful farming practices and uncontrolled urbanisation are major causes of environmental degradation.

Carbon dioxide emissions are widely believed to be a major source of global climate change, and the burning of fossil fuels is their main source. The one-fifth of the world's population, living in the industrialised countries, accounts for nearly 60 per cent of the world's total consumption of energy, but the developing world's share is rising rapidly.

The world has made progress since 1972, when the United Nations convened the first global conference ever to address environmental issues. That conference stimulated the creation of environmental ministries throughout the world, established the United Nations Environment Programme and led to a vast increase in the number of civil society organisations promoting environmental concerns.

Twenty years later, in 1992, the United Nations' Conference on Environment and Development, (UNCED) provided the foundation for agreements on climate change, water bodies and biodiversity. It adopted an indicative Policy Framework intended to help achieve the goal of sustainable development — in rich and poor countries alike.

Perhaps the single most successful international environmental agreement to date has been the Montreal Protocol, in which states accepted the need to phase out the use of ozone-depleting substances.

The Millennium Report of the Secretary General of the United Nations has pointed out that, "Nevertheless, we must face up to an inescapable reality: The challenges of sustainability simply overwhelm the adequacy of our responses. With some honourable exceptions, our responses are too few, too little and too late."

Major Challenges for Sustainability

The four major challenges for sustainability have been summarised in the Millennium Report, as follows:

(i) Coping with climate change.
(ii) Confronting the water crisis.
(iii) Defending the soil.
(iv) Preserving forests, fisheries and biodiversity.

Coping with Climate Change: The increasing carbon emissions over the past half-century alone have resulted in warming of the earth's atmosphere at an increasing rate. The summer of 1998 was the hottest on record and the winter of 1999-2000 was referred as the warmest.

The cost of natural disasters in 1998 alone exceeded the cost of all such disasters in the entire decade of the 1980. Tens of thousands of mostly poor people were killed that year and an estimated 25 million environmental refugees were forced out of their homes.

The main cause for the threat of climate change is the burning of fossil fuel which provides more than 75 per cent of energy worldwide. The rapidly expanding number of automobiles around the globe, threaten an even greater escalation in emissions. The need to promote energy efficiency and greater reliance on renewable resources is obvious.

Stabilising levels of carbon dioxide to a range that is considered safe will require overall reduction on the order of 60 per cent or more in the emission of the greenhouse gases that are responsible for global warming.

Confronting the Water Crisis: Global fresh water consumption rose 6-fold between 1900 and 1995 — more than twice the rate of population growth. About one-third of the world's population already lives in countries considered to be water stressed — that is, where consumption exceeds 10 per cent of total supply. If present trends continue, two out of every three people on earth will live in that condition by 2025.

Ground water supplies to about one-third of the world population, the unsustainable but largely unnoticed, exploitation of this water resource is a particular cause of concern. In a world where 30-40 per cent of food production comes from agricultural lands, this is a critical issue for food security.

The most serious challenge is the fact that more than 1 billion people lack access to safe drinking water, while half of humanity (3 billion people) lacks adequate sanitation. In many developing countries, rivers downstream from large cities are little cleaner than open sewers. Unsafe and poor sanitation cause an estimated 80 per cent of all diseases in the developing world. The annual death toll exceeds 500 million.

Defending the Soils: Increasing land degradation reflected in the 2 billion hectares of land is affected by human induced, degradation of soil, putting the livelihoods of nearly 1 billion people at risk. The major culprits are, irrigation induced, salinisation, soil erosion caused by overgrazing and deforestation and biodiversity depletion. Each year an additional 20 million hectares of agricultural land becomes too degraded for crop production.

The world faces a real threat to future global food security. Advances in agriculture biotechnology may help developing countries by creating drought, salt and pest resistant crop varieties. But, the environmental impact of bio-technology has yet to be fully evaluated and many questions, in particular those related to bio-safety, remain to be answered.

Preserving Forests, Fisheries and Biodiversity: Some 65 million hectares of forests were lost in the developing world during 1990-1995 because of over-harvesting, conversion into agricultural land, disease and fire. Biodiversity is not only important for agriculture; plant based medicines provide more than 3 billion people with their primary health care and comprise a multi-billion dollar a year global industry. The survival of some 25 per cent of the world's mammal species, and 11 per cent of bird species is also threatened. Ocean fisheries continue to be stressed despite the large number of regulatory agreements in place.

Globalisation, Economic Development Approach and Environmental Degradation in Asia — The Poorest Region in the World

At present, development strategies in the Asia region are heavily focused on fostering economic growth, using the private sector as the primary vehicles for undertaking the related economic activities. Such emphasis often does not pay due attention to the detrimental environmental effects of the modalities used to implement the strategy. Increasingly, the need to protect the fragile environment, and thus make economic development sustainable in the full sense of the term, is receiving the attention of policy makers.

Despite domestic and international efforts, Asia continues to be the most polluted and environmentally degraded region while the Pacific region faces the problem of diminishing natural resources and other sustainable development concerns. The range of problems are huge — from degradation of rural land to the pollution and congestion of the urban areas — the so-called Mega Cities. Some areas have unique problems of sea level rise (Pacific). Some environmental problems are local in nature — such as air and water pollution. Based on the Emerging Asia: Changes and Challenges, the ADB Study and the State of the Environment in Asia and the Pacific, 2000 (published by ESCAP and ADB), some of the critical situation relating to land, forest, air quality water, quality and solid waste are stated as follows:

Land

- Less land per capita (0.3 hectares compared to 1.6 in the rest of developed countries)
- Food production has not matched the population growth rate
- Soil erosion, soil salinisation, desertification continue unabated
- Agricultural intensification are causing health problems: 27 per cent deaths are related to environmental causes.

Forests

- Shrinking by 1.2 per cent per year
- Lost almost half in the past 30 years

Air Quality

- 10 of Asia's 11 mega cities exceed WHO guidelines of particulates by factor of at least 3; four countries exceed acceptable lead levels, three exceed acceptable ozone and sulphur dioxide levels.
- 13 of the 15 dirtiest cities are in Asia.

Water Quality

- Suspended particulate highest in the world (due to deforestation, unsustainable watershed management and agricultural practices)
- Organic wastes (BOD) — 1.4 times world average
- Contains 3 times as much bacteria (fecal coliform) than world average
- The severity varies within Asia, i.e. in Beijing and Manila only 15 per cent waste water is treated. These are the best examples in the region

Solid and Hazardous Wastes

- A study shows 1 per cent income rise, causes 0.34 per cent rise in municipal solid waste
- On average municipalities spend 50-70 per cent of their revenues on management. Still collection services are low: 50-70 per cent at the most
- Hazards and toxic wastes are on the rise with new petrochemical industries coming up (The growth and affluence brings the problems of hazardous wastes)

Environmental neglect has had a profound effect on economic development. A recent study has revealed that in China, the environmental damage owing to productivity losses from land degradation, water scarcity and destruction of wetlands amounted to 3.8-7.3 pe rcent of GNP in 1990, which accrues to US$ 13.9 to 26.6 billion. In Thailand, the health cost related to particulate and lead contributed to a loss of 2 per cent of GNP. Similar situations are seen in Indonesia, Pakistan and the Philippines ranging from 0.8 to 3.3 per cent of GNP.

The Future Course of Action and Priorities

The Secretary General, United Nations has specified the need to build new ethics of global stewardship. It would be appropriate to consider those elements. He specified certain major causes of priority actions. The present day ecological crisis encompass many of the following causes:

(i) Poverty.
(ii) Negligence.
(iii) Greed.
(iv) Failures of Governance.

These crises do not admit of easy or uniform solutions.

Moreover, there is every reason to expect that unpleasant ecological surprises lie ahead. It is worth recalling that neither global warming nor ozone depletion were on the agenda of the 1972 United Nations Conference on the Human Environment in Stockholm. Nor would anyone in 1970 have predicted that the cost of natural disasters would increase 900 per cent between the 1960s and 1990s.

It is true that technological breakthroughs that are unimaginable today may solve some of the environmental challenges we confront. Perhaps they will, and we should surely provide incentives to increase the likelihood of their occurring. But it would be foolish to count on them and to continue with business as usual.

The priorities can be expressed as follows:

First, major efforts in public education are needed. Real understanding of the challenges we face is alarmingly low. As more and more of us live in cities, insulated from nature, the need for greater awareness grows. Consumers everywhere have to understand that their choices often have significant environmental consequences.

Much of the burden of consciousness-raising to date has fallen on civil society organisations. With energy, commitment, but few resources, non-governmental organisations have advocated environmental issues in public debates almost everywhere. Schools and universities also have a critical role to play in raising public consciousness, and governments themselves must step up their contributions.

Second, environmental issues must be fundamentally repositioned in the policy-making process. Governments typically treat the environment as an isolated category, assigned to a relatively junior ministry. This is a major obstacle to achieving sustainable development. Instead, the environment must become better integrated into mainstream economic policy. The surest way to achieve that goal is to modify systems of national accounts so that they begin to reflect true environmental costs and benefits — to move towards "green" accounting.

Today, when factories produce goods but in the process pump pollutants into rivers or the atmosphere, national accounts measure the value of the goods but not the costs inflicted by the pollutants. In the long run, these unmeasured costs may greatly exceed the measured short-term benefits. Only when they reflect a fuller accounting can economic policies ensure that development is sustainable.

The system of Integrated Environmental and Economic Accounting, pioneered by the United Nations in 1993, is a response to this challenge. It augments traditional national accounts with natural resource and pollution flows accounts. This additional information enables governments to formulate and monitor economic policies more effectively, enact more effective environmental regulations and resource management strategies, and use taxes and subsidies more efficiently.

Although this system of green accounting is still a work in progress, it is already employed by national governments. The Government of the Philippines started using it in 1995. Another 20 or so countries. North and south, are using elements of it.

Third, only governments can create and enforce environmental regulations, and devise more environment-friendly incentives for markets to respond to. To cite but one example, governments can make markets work for the environment by cutting the hundreds of billions of dollars that subsidize environmentally harmful activities each and every year. Another is by making greater use of "green taxes", based on the "polluter pays" principle.

Creating new incentives also encourages the emergence of entirely new industries, devoted to achieving greater energy efficiency and other environment-friendly practices. The success of the Montreal Protocol, for instance, has created a large market for ozone-safe refrigerators and air conditioners. Nothing would be more foolish than neglecting the enormously positive role the private sector can play in promoting environmental change.

Finally, it is impossible to devise effective environmental policy unless it is based on sound scientific information. While major advances in data collection have been made in many areas, large gaps in our knowledge remain. In particular, there has never been a comprehensive global assessment of the world's major ecosystems. The planned Millennium Ecosystem Assessment, a major international collaborative effort to map the health of our planet, is a response to this need. It is supported by many governments, as well as UNEP, UNDP, FAO and UNESCO.

Conclusion

The poor and developing countries need economic development at a fast pace. Whether this development process is sustainable or not depends on governments' specific interventions in terms of policies conducive to environment preservation; growth of necessary institutions and public involvement in the process of development.

The guidelines laid down by the Secretary General, United Nations are indicator enough for countries to decide severally their approach ; and the global community to decide their collective approach.

Globalisation will best serve the economic development objective when strategies to balance environmental concerns with the promotion of economic growth become a major aspect of government policies and plans, although those strategies are constrained by a combination of factors. Integration of environmental considerations into economic policy decisions requires:

(i) Institutional framework to facilitate such integrated approach;
(ii) Effective communication between the environment and economic policy makers to enable meaningful policy making; as well as;
(iii) Effective policy tools which lead economic actors to incorporate environmental factors into determinants of their economic decision making.

The future globalisation has to be in the form of 'inclusive globalisation' so that the poor have an equal opportunity to be benefited.

23

Sustainable Development Needs New Focus

India has faced the decade of 90s introducing the process of liberalisation and globalisation of economy. This change was considered not desirable by the traditional industrialists, but a challenge by the entrepreneurs. Moving from planned and administered economy to liberalised economy has been a difficult task. To understand the difficulty in transition it is necessary to understand the basic difference in approach under both the regimes.

The critical difference between planning and liberalisation is that planning assumes that outcomes are predictable, while liberalisation facilitates the unpredictable. It does so by unshackling the energy of innovation of millions of people unknown to planners, of letting them play around with ideas unthought of. Such unknown people and ideas have transformed other countries and will also trans from India.

A look at the market capitalisation in 2000 reflects that dynamic Indian companies are now leading the market. The market capitalisation (market value of stocks) of WIPRO (Rs. 68,000 crores) and Infosys (Rs 51,000 crores) have overtaken that of the traditional MNC leader Hindustan Lever (Rs. 48,000 crores).Other upstarts in the information industry like NIIT (Rs. 12,700 crores). Satyam (Rs. 6,300 crores) and Pentafour Software (Rs. 2,800 crores) have overtaken long-entrenched MNCs in India like ICI (Rs. 1,000 crores), Siemens (Rs. 1,400 crores) and Proctor & Gamble (Rs. 1,500 crores). Traditional Indian business class Tatas, Birlas and Modis find it difficult to fight the MNCs. The understanding was that these big businesses house dictated the future trebds in Indian business as leaders. The effect of liberalisation and

---

Paper presented at the Round Table on Sustainable Development organised by South Asia Foundation on June 5, 2001 at New Delhi.

globalisation has made it amply clear that the talent do not lie with the traditional industries, but with a large number of independent unknown people who had a better understanding of the changing scenario. They realised that the next age is of information technology and the knowledge based systems will rule the market. It is not to say that IT companies as software and hardware producers are required. But, that IT interventions will be necessary for future development. What is called for is that trained manpower in every sector should have sufficient knowledge of 'knowledge based systems'.

India is striving hard to achieve sustainable development. Certain elements that affect sustainability are being discussed first and then the subsequent need of new directions in skills building.

Elements Crucial for Sustainable Development

The UNEP report, Global Environment Outlook-1 (1997) mentions that significant progress has been made in the last decade in confronting environmental challenges in both developing and industrial regions. Worldwide, the greatest progress in the realm of institutional developments, international cooperation, public participation, and the emergence of private sector action, legal frameworks, economic instruments, environmentally sound technology, and cleaner production processes have been developed and applied. Environmental impact assessments have become standard tools for the initiation, implementation, and evaluation of major development and investment projects in many countries around the world.

As a result, several country reports marked progress in curbing environmental pollution and slowing the rate of resource degradation as well as reducing the intensity of resource use. The rate of environmental degradation in several developing countries has been slower than that experienced by industrial countries when they were at a similar stage of economic development.

Despite this progress on several fronts, from a global perspective, the environment has continued to degrade during the past decade, and significant environmental problems remain deeply embedded in the socio-economic fabric of nations in all regions.

Although poverty and the growing global population are often targeted as responsible for much of the degradation of the world's resources, other factors — such as the inefficient use of resources, waste generation, pollution from industry, and wasteful consumption patterns — are equally driving us towards an environmental precipice. In many instances, although trends are increasing, the rate of increase over the years has slowed down or was less than the rate of increase in economic growth previously experienced by countries with comparable economic growth. This suggests that several countries are making the transition to a more sustainable environment at a lower level of economic development than industrial countries typically did over the last 50 years.

Several issues that need to be considered for sustainable development include the following:

(i) Population
(ii) Hunger and Malnutrition
(iii) Poverty
(iv) Unemployment
(v) Water
(vi) Air

Where do we go?

All these issues lend to one basic question.

Environmentally Sound Technologies (ESTs) hold the Key for Future Economic Development

Agenda-21 emphasises the importance of environmentally sound technologies (ESTs). They "protect the environment, are less polluting, use all resources in a more sustainable manner, recycle more of their wastes and products, and handle residual wastes in a more acceptable manner than technologies for which they are substitutes". Agenda-21 also called on governments and other players to develop new funding mechanisms to accelerate the transfer of ESTs from the 'haves', mainly in the industrialised countries, to the 'have-nots' chiefly in the developing world. And it underlined the point by urging business leaders to give environmental management " the highest priority" in the move towards sustainable industrial practices, calling on business and industry to "develop techniques and technologies that reduce harmful environmental impacts". Agenda-21 also stressed that "new and efficient technologies will be essential to increase the capabilities, in particular of developing countries, to achieve sustainable development, sustain the world's economy, protect the environment, and alleviate poverty and human suffering".

What are ESTs? They cover a wide range of process and product technologies. Some, such as solar energy photovoltaics, fibre-optic devices, electric or ultra fuel-efficient vehicles, and biotechnology-based solutions, are available now, but are not yet economical enough to be commercially applied on a wide scale. In time, these and other emerging technologies will transform both production and consumption patterns, drive much of industry towards sustainable production, and eventually move some industrial sectors to the goal of zero emissions.

But the crucial point is that, meanwhile, many other ESTs, both 'end-of-pipe' and 'prevention technologies', are already available to achieve significant, sometimes dramatic, environmental improvements. Cleaner production technologies focus on achieving process and product change both to reduce pollution and also to prevent it occurring in the first place. End-of-pipe ESTs are more limited in scope, but they do control and reduce waste and emissions, as well as clean up pollution after it has happened. There is now a paradigm shift taking place towards the cleaner production approach. As Agenda-21 makes clear, ESTs "are not just individual technologies, but total systems which include know-how, procedures, goods and services, and equipment as well as organisational and managerial procedures".

Industry, worldwide, though mainly in the industrialised countries, is using ESTs on a wider scale. One measure — and it is only one measure — is that the global market for environmental products such as water and wastewater treatment, waste management and recycling, air pollution control, noise reduction systems and services, is growing fast. It is increasing by around 5 per cent every year and is set to reach, conservatively, an annual total of US$300 billion by the year 2000. The United Nations Development Programme (UNDP) projects the global compliance market (including ESTs required to meet mandatory environmental standards) to reach US$500 billion a year by 2000. It forecasts that the future global market for environmental investments will expand to US$300-600 billion annually for pollution control, a total of US$1 trillion for power from 1993 to 2000, and a total of US$250 billion for energy conservation over the next 20 years.

What is difficult to calculate from these figures is the split between end-of-pipe and cleaner production technologies, although at the moment the market for the later is certainly much smaller. While end-of-pipe ESTs do help to reduce emissions, they are transitional rather than higher-term technology and should be used in conjunction with cleaner production approaches. Eventually, cleaner production process should take over and completely replace end-of-pipe technologies.

Nonetheless, the market growth in ESTs does show that industry is moving to tackle environmental pollution problems.

There are several pressing requirements if further progress is to be achieved:

- to develop and commercialize new, advanced ESTs;
- to continue the shift from end-of-pipe approaches to cleaner production and eco-efficiency — that is, to move faster from pollution control to pollution prevention;
- to accelerate the knowledge and adoption of both new and existing ESTs throughout industry, in the industrialised world and, in particular, in the developing countries;
- to make the policy changes that will support the wider use of ESTs at the national and local level; and
- to support the development of local ESTs.

As the UNEP GEO-1 report states: "Appropriate technological improvements, which result in more effective use of natural resources, less waste, and fewer pollutant by-products, are required in all economic sectors. Truly global availability and worldwide application of best available and appropriate technology and production processes need to be ensured through the exchange and dissemination of know-how, skills and technology, and through appropriate finance mechanisms."

The UNEP report highlighted three main barriers to accelerating the use of ESTs:

(i) lack of knowledge about the existence of commercially available technologies and the benefit they bring;

(ii) lack of financial resources to get them where they are needed; and

(iii) lack of political action to create the right framework conditions for industry. While some governments are now using a mix of command-and-control policies and market-based incentives to push industry faster towards cleaner, more resource-efficient production, actual delivery on the political commitments made in Rio has, generally, fallen far short of what was promised.

The first two barriers apply particularly to developing countries though they are also a concern to small and medium-sized enterprises (SMEs) in many industrialised economies. The third relates to both developing and industrialised countries because there is a direct cause and effect between government policies and industry actions to improve environmental performance. There is no doubting the crucial role that government has in encouraging a faster take-up of ESTs as part of its move to promote sustainable development.

Future Course of Action

It is amply clear that the future course of development will have two important elements that will dictate the kind of skills required and the orientation to engineering education:

(i) Sustainable development in India will require introduction of clean technologies and cleaning up technologies.

(ii) Economic development in India will require IT interventions in traditional and emerging industries.

Certainly this development orientation will call for revolutionary changes in course curricula of engineering faculties. It is not sufficient to have super speciality course in IT sector and environmental engineering, but to develop interactive course that will expose engineering students to IT and environmental engineering.

Conclusion

The above mentioned issues identify the course for sustainable development. There is no ideal single solution for different geographic locations. The basic question right now is 'as to how fast can we convince the governments and the people to change from current living practices to environment friendly practices'. It is possible to make fast progress by tackling today's environmental problems and following the path of sustainable development.

24

Corporates and Sustainability

Sustainability — The Main Concern

World Summit on Sustainable Development, at Johannesburg ended on September 4, 2002, emphasised that much had been achieved since Rio. New international legal instruments had been developed, awareness had increased and progress had been made at the local, national and international level in confronting environmental challenges and achieving sustainability. At the same time, new scientific evidence of the planetary dimensions of global environmental change had raised the need for a quantum increase in our efforts. The world leaders agreed that this was the Summit of implementation, the Summit of accountability and of partnership. These leaders thought that concrete implementation must be the focus of future work. However, the final political document reflected this agreement only in part, and without any quantifiable targets.

The UNEP Global Environment Outlook, compiled as a result of the work of hundreds of scientists around the world, had singled out the root causes of global environmental degradation. They were embedded in social and economic problems such as pervasive poverty, unsustainable patterns of consumption and production and vast and increasing inequities in wealth distribution. The world was characterised by divided and dysfunctional cities, dwindling water supplies and potential conflict over scarce resources and the accelerating loss of the environmental capital that underpinned life on Earth. Problems of planetary dimensions required global responses. Investing in sustainable development would be investing in the future security of all.

Paper presented at the World Congress on Corporate Governance organised by the World Council for Corporate Governance during September 27-29, 2002 at New Delhi.

Just as unsustainable patterns of development result in environmental degradation, negative environmental trends impacted most severely on the poor, on those most poorly equipped to deal with them and often those who had the least responsibility for causing them.

The challenge is great, but perhaps figures on the economic impact of protecting biodiversity and ensuring ecosystem services could provide impetus for action. It had been estimated that the value of biodiversity was some $3 trillion per year, with overall ecosystem services estimated at over $33 trillion. That figure was practically the gross national product (GNP) for the entire planet. While those figures could be debated, the extreme value of ensuring functioning ecosystems and the biodiversity interplay — at least in the economic sense — was undeniable.

The Simple Question

When agencies and organisations addressed the World Summit on Sustainable Development, delegations were told that the Summit had one simple question to answer:

> *How could the next 10 years be more successful in addressing the fight against poverty and environmental degradation than the 10 years since Rio?*

Linkage of Sustainability and Development — Some Institutional Observations

Thus, it was quite evident that the total concern was on sustainability of the planetary system. Different actors had their own role which was well defined in the statements issued by different organisations. Some of the relevant remarks are reproduced to give an idea about how the issue of production and consumption patterns, poverty alleviation, economic development and social development were linked to job led development where the corporates have to play an important role.

Gro Harlem Brundtland, Director-General of the World Health Organisation (WHO), told delegates that the world had yet to give sustainable development the priority it deserved. The voices that were faint in Rio were louder now and they must be heard.

People were at the centre of sustainable development and the current crisis in southern Africa was a reflection of insufficient investment in people's basic needs. The result was vulnerability to the effects of drought. There was illness — most of all HIV/AIDS — underfunded health services, insufficient clean water, environmental degradation and social systems that could not cope.

The notion of a people-centred approach to sustainable development was also echoed by *Mary Robinson*, United Nations High Commissioner for Human Rights. She said focusing on individuals and, in particular, their indivisible human rights, added value, providing a normative framework of obligations that had the legal power to render governments accountable. Thus, she continued, poverty was seen in terms of the individual's right to food and clean water, shelter, health and work opportunity. Empowerment of the poor flowed from the recognition that they experienced the non-fulfilment of their rights.

Mark Moody Stuart, Chairman, Business Action for Sustainable Development, said that business had come to Johannesburg with a commitment to work in partnership with other major groups, governments and agencies to deliver sustainable development. Perhaps the most important element needed to facilitate the growth of economic activity, which was essential to achieving sustainable development, was sound governance systems at a national and local level in each and every country.

By sound governance he meant institutions whereby all sectors of society felt that their views had been taken into account and that the outcomes were fair and equitable. That would include sound governance of business, with the rules and frameworks necessary for markets to operate fairly and openly in every country, and with appropriate environmental regulation applied impartially to all. However, it would also include such elements as the rule of law, security, human rights and intellectual property. The conditions created by sound local governance were not only beneficial to the environment and society, but were exactly the conditions in which sound economic activity flourished, he said. Also, investment would flow to areas where such sound governance was in place or developing. He was also committed to a number of initiatives, in partnership with governments and others, to increase investment and to grow business in developing and least developed countries. That included paying close attention to the growth of the small and medium enterprise sector, which was not, by and large, represented at the Summit and, yet, which was so essential to sound development.

Juan Somavia, Director of the International Labour Organisation

(ILO), said the Summit had one simple question to answer:

> *How in the next decade could the international community be more successful in terms of sustainable development, job creation, the fight against poverty and environmental degradation than it had been in the 10 years since Rio?*

First, it would be necessary to acknowledge that changing unsustainable production and consumption patterns, through environmental friendly technologies, meant a revolution in the way the international community worked and the things it made. The current generation would have to retool the entire economic system, particularly its fiscal policies.

While that would be a daunting challenge, it would also present a massive opportunity for technological breakthroughs, investment, skills development, gender equality and decent work. He said it was important to root such a strategy in the workplace — where many of the initiatives would ultimately succeed or fail. It was organised workers and employers that would play the primary role in making the technological transition to sustainability.

Marcel A. Boisard, Executive Director, United Nations Institute for Training and Research (UNITAR), said that sustainable development was the responsibility of each and every person. Capacity training and building was, therefore, of the utmost importance. UNITAR designed and conducted many programmes in natural resource management and the environment, among other areas. Institution-building and strengthening was the ultimate goal of training.

Everyone must be properly trained to ensure the right kind of development management, he said. Capacity-building focused on the need of human development, better management and good governance. Sustainable development was holistic and highly diversified. UNITAR was a small and autonomous body, which had responded flexibly to the changing training needs of its members. Many of its programmes dealt with sustainable development. Describing several of UNITAR's programmes, he said the organisation had addressed methodology as well, including the need for assessment, evaluation of outcome, and ongoing training.

Mohamed T. El-Ashry, Chief Executive Officer and Chairman, Global Environment Facility (GEF), said that the Facility was considered by many to be the only major financial accomplishment of

Rio. It had provided $4.2 billion in grants, while leveraging $12.4 billion in additional financing for the global environment. Among its accomplishments, the Facility had worked with large and small businesses to deliver $6 billion in new renewable and clean energy services to developing countries.

Trine Lise Sundness, Norwegian National Union Center, speaking on behalf of the International Confederation of Free Trade Unions, said more than 2 million people died annually because of occupational accidents and diseases, resulting from unsustainable development. That was why working people were interested in the issue.

Tjaco van den Hout, Secretary-General of the Permanent Court of Arbitration, said the international community and civil society were growing increasingly dissatisfied with procedural defects in available mechanisms for resolving environmental disputes. Much of that dissatisfaction focused on the lack of a unified forum to which all actors involved in an environmental or natural resource related dispute could turn to. With the notable exception of the International Tribunal for the Law of the Sea, there was a paucity of institutional forums for such disputes.

Jose Medem, President, World Federation of Engineering Organisations, said scientists and engineers were committed to sustainable development through knowledge and technology generation and diffusion. He said, "Addressing the basic needs in the areas of water, food, health, housing and energy consisted essentially in the application of appropriate technologies."

He went on to say that countries in both the developed and developing world contributed significantly to global risks, such as climate change, ozone depletion, degradation of biological resources and loss or deterioration of arable land. The unsustainable production and consumption patterns of the developed world were largely responsible for the present condition of the Earth. Unfortunately, many in the developing world did not benefit as the path to prosperity viewed that same intensive model of economic development. Improving the economic well-being of lesser developed countries and reducing poverty of the third world was a prerequisite for creating a stable sustainable world.

In order to strengthen the capability of developing countries to produce, adapt to and use new knowledge and technologies, the international community must improve the dissemination of information through networks and partnerships.

Corporate Governance and Political Declaration

The Political Declaration, which was the commitment of various world leaders at WSSD, included the following:

- We recognize that the process of globalisation is accompanied by the emergence of leading private sector corporations which have a responsibility to contribute to the evolution of equitable and sustainable communities and societies, even as they pursue their legitimate activities.
- We, therefore, agree that there is need for the private sector to operate within a transparent and stable regulatory environment to reinforce its corporate responsibility and social contribution.
- Accordingly, we agree that the United Nations General Assembly should pursue the matter of corporate responsibility and the social contribution of the private sector.

Globalisation has added a new dimension to the challenges of: poverty eradication; ever increasing gap between the rich and the poor; and deterioration of global environment. The rapid integration of markets, mobility of capital and significant increases in investment flows around the world has opened new challenges and opportunities for the pursuit of sustainable development. But the benefits and costs of globalisation are unevenly distributed, with developing countries facing special difficulties in meeting this challenge.

We risk the enrichment of these global disparities and unless we act in manner that fundamentally change their lives.

Plan of Implementation and Responsibilities of Corporates

The Plan of Implementation specifically calls for enhancement of corporate environmental performance through voluntary initiatives, including environmental management systems, codes of conduct, certification and public reporting on environmental and social issues, taking into account such initiatives as the International Organisation for Standardisation (ISO) standards and Global Reporting Initiative guidelines on sustainability reporting, bearing in mind principle 11 of the Rio Declaration on Environment and Development.

- Encourage dialogue between enterprises and the communities in which they operate and other stakeholders.
- Encourage financial institutions to incorporate sustainable development considerations into their decision-making processes.
- Develop workplace-based partnerships and programmes, including training and education programmes.

The Plan also encourages relevant authorities at all levels to take sustainable development considerations into account in decision making, including national and local development planning, investment and infrastructure, business development and public procurement. This would include:

- Provide support for the development of sustainable development strategies and programmes, including in decision-making on investment in infrastructure and business development;
- Continue to promote the internalisation of environmental costs and the use of economic instruments, taking into account the approach that the polluter should, in principle, bear the costs of pollution, with due regard to the public interest and without distorting international trade and investment;
- Promote public procurement policies that encourage development and diffusion of environmentally sound goods and services;
- Provide capacity-building and training to assist relevant authorities with regard to the implementation of the initiatives listed in the present paragraph; and
- Use environmental impact assessment procedures.

Conclusion

Thus, it will be seen that environmental sustainability has a direct bearing on the role of corporates at the policy level as well as the implementation. An agreement on the political declaration at the WSSD is a solemn commitment of world leaders towards supporting the sustainability movement. The corporates in each country have now to play their role in discharging equal and differentiated responsibilities.

The World Environment Foundation in its Declaration and the Environment Management Congress at Palampur in June 2002 launched India Sustainability Movement, that clearly defines the future course of action. As the Vice-Chairman of the World Environment Foundation has strongly recommended that the participants to this Conference commit themselves to the Sustainability Movement.

25

Environmental Concerns for Effective Corporate Planning

Introduction

"Sustainable development" is not just a buzzword. To novice in the field of environment or to be precise, in economic development sector, it appears that any mention of "sustainable development" is sufficient to deal with modern business ethics. Sustainable development has one specific dimension for biological sectors and another for the non-living sector. The difference in approach is too fine to determine as to which one is true. In the ultimate analysis it is sustainability of the global systems that is paramount.

Environmental Concerns in the Industry — Status

Environmental concerns among a small group of large companies is growing in number. For most companies dealing with the environment is still primarily about compliance — a costly, cumbersome and unwelcome process. Environmental concerns, however, shape every aspect of strategy and action.

The chemical industry was the first to be caught in the controversy of environmental pollution in the 50s. The instantaneous response was to deny everything. But there was no reliable data. Government and the private industry were projecting their data to substantiate the respective argument. It took a long time for the industry and the government to establish a dialogue. Once the dialogue took place the chemical industry was quick to respond, which was beyond 'compliance alone' and addressed the global environmental issues on 'pro-active'

Paper presented at the World Congress on Corporate Governance organised by the World Council of Corporate Governance during 21-23 January, 2002 at Bombay.

to the 'voluntary' basis and this obviously meant emphasis on delivery of action, required to protect the environment. Subsequent global industrial activities reflected that responsible care and collective action clearly makes sense for a lot of reasons including, sharing the burden, the risk of being harmed by others' actions, and so on.

The industry also realised that courting long term prosperity through green marketing, or through claims that are not sustainable, is asking for trouble.

The industry has to set its priorities within the environmental arena. It is really a struggle because suddenly the environmental compliance is thrust by the government for which the industry is not prepared. A major multinational Dow Chemical Company has been spending over US$ 500 million per year on environmental compliance. The company has also set aside a budget for voluntary pre-emptive initiatives under what is called the Waste Reduction Always Pays Programme (WRAP). It has shown a 55 per cent return on investment and therefore, the company has appreciated what the environment is all about. The bottom line is that as an operating principle one cannot calculate the return on investment on an environment impact alone. A return on investment on the entire facility is carried out and that is compared to what the return on investment would look like without the environmental features. If a company were to work alone acting only in response to government pressure, it can be seen as self-serving and defensive. But, most of the global companies now have the courage, voluntarily to mandate for themselves global environmental standards. And, that is a good sign.

Major industries have also indicated that they are perfectly capable of preparing model laws and regulations, usually on a national basis, in response to a threat, but that is too late. These companies have seriously to think if they can put them together in advance of the public outcry, so that they are seen as the solution rather then the problem. The industry needs to work with governments to get pollution prevention instilled in manufacturing, instead of resorting to end-of-the-pipe treatment. The industries can then make money on it. This is what the Business Council for Sustainable Development called 'eco-efficiency' — 'economic and environmental efficiency'.

When an assessment is made, and corporate decisions are taken on capital expenditures, it becomes necessary to identify the additional capital that has to be spent on a plant to achieve the degree of environmental protection being aimed for, and then assess whether that piece of capital is going to add or subtract from shareholder value in the long term. Some major companies have, in the past, abandoned projects after doing this analysis. They have chosen not to build a plant at all because they could not afford to put in as much environmental protection as world wanted.

Another important input for corporate planning is to develop a body of knowledge that gives a fair valuation to key raw materials and energy. If someone knows, as to what it cost to grow a tree or to process a barrel of oil, to generate a calorie or a BTU or whatever, then he can build a bridge between pollution reduction and pollution prevention. It is an issue that has been widely discussed and that takes to full cost accounting. The industry has realised that the problem with current accounting is that people do not allocate environment cost to their products. Many companies still assume, they can run incrementally, dealing with today's issues, one piece at a time with no thought for the big picture. They build mountains of wastes, they exploit what they see as competitive advantages between various materials. In the end it means that by failing to account for the full cycle of production, use and disposal, they often subsidies the wrong choices.

Full-cost accounting is simply about allocating the full cost—including environmental cost that arise during and disposal—to a given product or process. The reality is that many a times, industry end up making a wrong choice of product in ignorance.

Environmental Damage and Growing Use of Environmentally Sound Technologies (ESTs)

It is difficult to make a specific assessment. However the UNDP tried in its Human Development Report, 1998 to reflect environmental damage cost to natural resources in a few selected countries. Various estimates are only indicative. These highlight the possible dimension of the problem, but does not exactly quantify the problem.

While the situation is grim, it does provide the opportunity to introduce environmentally sound technology (EST) solutions on a massive scale. The enormity of the damage caused is reflected in the table below. To counter it the enormity of the need is clear. For example that for China it has been estimated that an investment of US $ 50 billion is needed in technologies to mitigate greenhouse gas and acid rain emissions. Such investment is already beginning to happen. All these investments are being made under the GEF or through the commercial routes. In every case it is the investment through the industry alone. China is investing at a rate of about US $ 2 billion per year for clean technologies.

Situation in India is no different. With the economic reforms and impact of globalisation the industry is modernising very fast. They have realised the need for clean and cleaner technologies for production.

Table 25.1: Estimated Cost of Environment Degradation in Selected Asian Countries

Country	*Year of Period*	*Environmental Damage*	*Annual cost (US$ billion)*	*Cost as % of GDP (per cent)*
1. China	1990	*Productivity losses caused by soil erosion, deforestation and land degradation, water shortage and destruction of wetlands. *Health and productivity losses caused by environmental pollution in cities.	13.9-26.6	3.8-7.3
2. Indonesia	1989	*Health effects of particulate and lead levels above WHO standards in Jakarta.	6.3-9.3	1.7-2.5
3. Pakistan	Early 1990s	*Health impacts of air and water pollution and productivity losses from deforestation and soil erosion.	1.7	3.3
4. Philippines	Early 1990s	*Health and productivity losses from air and water pollution in the vicinity of Manila.	0.3-0.4	0.8-1.0
5. Thailand	1989	*Health effects of particulate and lead levels above WHO standards.	1.6	2.0
6. India	1992	*Costs of urban air pollution, water pollution, soil erosion, land degradation and deforestation.	10-13.8	4.5-5.0

Source: Human Development Report, 1998.

Industry worldwide, though mainly in the industrialised countries, is using ESTs on a wider scale. One measure – and it is only one measure – is that the global market for environmental products such as water and wastewater treatment, waste management and recycling, air pollution control, noise reduction systems and services, is growing fast. It is increasing by around 5 per cent every year and is set to reach, conservatively, an annual total of US$ 300 billion by the year 2000. The United Nations Development Programme (UNDP) projects the global compliance market (including ESTs required to meet mandatory environmental standards) to reach US$ 500 billion a year by 2000. It forecasts that the future global market for environmental investments will expend to the extent US $ 300-600 billion annually for pollution control, a total of US$ 1 trillion for power from 1993 to 2000, and a total of US$ 250 billion for energy conservation over the next 20 years.

What is difficult to calculate from these figures is the split between end-of-pipe and cleaner production technologies, although at the moment the market for the later is certainly much smaller. While end-of-pipe ESTs do help to reduce emissions, they are transitional rather than higher-term technology and should be used in conjunction with cleaner production approaches. Eventually, cleaner production process should take over and completely replace end-of-pipe technologies. Nonetheless, the market growth in ESTs does show that industry is moving to tackle environmental pollution problems.

The market for ESTs has grown for two main reasons:

- Companies have been required to meet increasingly strict regulatory rules, national regulations and international treaties for pollution and waste reduction. For many companies, the first step has been to use end-of-pipe technologies, which is a short-term and expensive solution. The ultimate goal should be to move beyond this and use cleaner production processes.
- Many corporate leaders, accepting that sustainable development is an integral part of the business agenda, have moved proactively to improve their companies' environmental performance without waiting to be mandated to do so. More and more companies have recognised that minimising, better still eliminating, undesirable harmful effluents, emissions and wastes, improves their performance both environmentally and economically. An emphasis on technological innovation has been a feature of this approach.

This last point is proving an increasingly important factor, for there is abundant evidence of the considerable financial benefits from using ESTs from companies themselves. The return on the investment has been significant and, as the World Business Council for Sustainable Development says, companies that introduce new technology to improve environmental performance before mandated to do so by regulations, cut costs and boost competitiveness.

Impressive Results

Thanks to ESTs, industry has been improving and continues to improve its environmental performance. For example, the industrialised countries have achieved major environmental quality gains during the past 20-25 years at a time when their economies have grown by 80 per cent. The industrialised countries account for 80 per cent of the market for environmental products. The cost has been 0.8 to 1.5 per cent of gross domestic product (GDP), borne equally by the private and public sectors. The fact that many environmental problems remain, and in some cases are worsening, takes nothing away from what the industry has achieved, but it does not underline how much more still has to be done.

The Organisation for Economic Cooperation and Development (OECD) says that one lesson from the progress achieved is that "many of the environmentally sound technologies and practices developed in the industrialised countries can be adapted to the needs of the developing countries". They need to be because achieving sustainable economic growth in the developing countries poses an increasingly formidable challenge.

Main Needs

There are several pressing requirements if further progress is to be achieved:

- to develop and commercialize new, advanced ESTs;
- to continue the shift from end-of-pipe approaches to cleaner production and eco-efficiency — that is, to move faster from pollution control to pollution prevention;
- to accelerate the knowledge and adoption of both new and existing ESTs throughout industry, in the industrialised world

and, in particular, in the developing countries;

- to make the policy changes that will support the wider use of ESTs at the national and local levels;
- to support the development of local ESTs.

Environment Issues in Corporate Decision-Making

The synergy between business and environment calls for a self-supporting market mechanism working independent of government's props like subsidy. The synergy cannot be effected unless appropriate elements are built into the environment policy and the economic calculus of industry and business on a reasonable time span basis of sustainability. Budget speech of the Finance Minister of India for 1998-99 also reflected the need for environment-friendly business growth. Though the NGOs at WTO meet at Seattle in November-December, 1999 opposed non-tariff barriers based on environmental and human rights issues, they never opposed environment-friendly business growth.

Various environmental issues important for corporate consideration are as follows:

Market Based Instruments for Environmental Management

The government has considered certain economic instruments that can correct market distortions. The general understanding is that the price will correct the current consumer behaviour and minimise work; and business will be working under true market mechanism. Certain considerations include the following:

- As against the Command and Control approach to management of the environment, the Economic or Market Based Instruments (MBIs) approach sends economic signals to the polluters to modify their behaviour. The approach normally involves financial transfers between polluters and the community and affects relative prices. But the polluters have freedom to respond and adjust, in the manner they want. They can, thus, choose the least cost option to meet the requirements. Hence, it is considered to be an efficient approach compared to the approach based on standards and regulations. The MBIs, therefore, have the benefit of being flexible and cost effective providing incentives for dynamic efficiency and resource transfer.

- Economic instruments used for environmental tax include pollution charges (emission/affluent tax/pollution taxes), marketable permits, deposit refund system, input taxes/product charges, different tax rates and user administrative charges and subsidies for pollution abatement. These can be both price based and quality based instruments.
- MBIs have been applied in both developed as well as developing countries. In general, price based MBIs have been more widely used than those which are quantity based. Within price based MBIs, indirect instruments such as input-output taxes, differential tax rates and user fees have found extensive application in developed countries. By contrast, developing countries have made greater use of subsidies including those for end-of-pipe treatment equipment.
- The main MBIs used in India are subsidies for pollution abatement equipment for air and water resources. This provides rebates on duties for various pollution control equipment, monitoring instruments and abatement machinery for air/water pollution and promotion of unleaded fuel/fuel efficient automobile subsidy on automobile pollution kits/ converters etc. Accelerated depreciation for pollution control machinery is also provided. Among user charges/ administration charges, consent fee is charged from industries under the Water Act and the Air Act. A water cess based on the consumption of water and type of industry (polluting) is also levied on selected industries and urban municipalities to conserve consumption and control pollution of water.

National Development Perspective — Consumption versus Preservation of Environmental Resource

Sustainable development is very much concerned with the availability of environmental and natural resources in the future. It goes beyond the static framework of correcting environmental externalities, deficient markets and inadequate property rights. The problem is much more acute in developing countries, like India, that have a high discount rate due to the scarcity of capital and prevalence of poverty (which indicates a strong preference for present over the future consumption). Use of appropriate discount rates/shadow prices is central to maintaining an optimal balance between consumption and preservation of natural resources. This approach is as crucial for the industry to adopt as for the government to initiate and maintain in its planning. This can be the starting point for synergy between business and environment.

Valuation of Environmental Damage

Attempts to establish optimal pollution abatement levels or to set more general environmental quality objectives of a cleaner to clean production confront several problems in determining the monetary value of environmental damage. Some of the reasons for the same are as follows:

(a) Firstly, for setting optimal abatement levels the economic activity must be functionally related to measurable environmental changes (and the scientific basis for establishing the linkages should be ensured).
(b) Secondly, only a use value of a good is normally accepted as contributing to welfare. Environmental goods have non-use value and option value (willing to pay for preserving the option to use the goods in future).
(c) Thirdly, many environmental services do not pass through markets and, therefore, no market prices exist to indicate value.

The corporate plans need to find a suitable measurable/quantifiable mechanism to assess the environmental damage cost of business activity and internalise that cost into pricing of product. Any abatement cost should also be considered and internalised.

Natural Resource Accounting

The use of GNP or GDP alone as an indicator of human welfare is no more considered sufficient. Accounting schemes that more accurately measure the environmental cost of economic activities at all levels could help increase the efficiency of natural resource use and reduce related environmental impacts. The need for natural resource accounting, NRI, and their integration with the system of national accounts have been emphasised.

Integrated environment and economic accounting, therefore, attempts at accounting for both, socio-economic performance and its environmental effects and integrating environmental concerns into main stream economic planning and policies. Given the experimental nature of some of the proposed methodologies, particularly those on monetary valuation of non-marketed assets and externalities, such environmental accounting would require numerous and controversial estimates and valuations. Larger industries have to play an important role in defining those interventions.

The Driving Forces

India is peculiar in that it has the characteristics of the developed economies as well as the developing economies. Therefore, the driving forces include the following elements:

- Pressure from central government, including increasingly stringent regulations.
- Improved enforcement of environmental regulations by newly empowered and resources policing agencies.
- Public pressure.
- The corporate environmental policies and programmes of the larger international companies who are forming new alliances.
- International donor assisted initiatives in environmental protection.
- The emergence of ISO 14001.

Certain Basic Queries Leading to Corporate Issues

The Corporate strategy may have to critically analyse some of the inputs by developing answers to the following queries:

- What role does the Corporate Strategy play in finding environmental solutions?
- Within the environmental arena how are the priorities decided?
- What kind of a linkage has been developed between the industry and the government to respond to environmental

regulation compliance?
- Can Market Based Instruments alone take care of the environmental concerns?
- How does the industry define environmental cost? What role does waste reduction programme play in environmental cost?
- How does the industry audit total spending on the environment?
- How are managers evaluated on the environment?
- How much time does the Chief Executive spend on environmental issues?

Conclusion

Globalisation of Indian economy has brought about consciousness in the Indian industry towards better quality and environmental measures. The industry is still enamoured by the end-of-the pipe technological options as those are cheaper to introduce. However a select group of industries are convinced for project based environmentally sound technology interventions. This has to be formalised so that regular production includes this in routine. The wider use of existing environmentally sound technologies would bring major environmental and economic gains, particularly in the shorter term. This benefit can accrue only through appropriate corporate strategies. The concerns mentioned above along with the driving forces can lead to reasonable approach for environmentally clean sustainable development.

The question is how fast will the situation change? It is perfectly possible to make much faster progress in tackling today's environmental problems: the technologies exist already — the challenge is to remove obstacles impending their use and industry making best of the opportunity.

REFERENCES

Chaturvedi, Pradeep. (1999). *et al. Business in a Cleaner Environment — The Indian Scenario*, World Environment Foundation.

Chaturvedi, Pradeep. (2000). *World Congress on Sustainable Development*. World Federation of Engineering Organisations. January 2000. Calcutta.

Visions for the 21st Century. Deutschland. December. 1999.

UNEP Report. *Sustainable Business — Economic Development and Environmentally Sound Technologies*. 1998.

26

Indian Industry Leads in Environment Protection

Introduction

'Environment Protection' has gained a significant place in the global development process. Once a much feared term in Indian Industry, it has become an acceptable entity. Involvement of the industry as a major stakeholder has allowed more financial, human and technical resources to be available to support the inter-governmental initiatives in environment cleanliness. Moving from 'end-of-the-pipe' efforts at the plant level to 'clean technologies' upfront has been a major achievement of the industry worldwide and in India. The pace of introduction of cleaner and cleaning-up technologies in India much higher than the pace anywhere else in the world. One must also appreciate that most of these achievements were attained at a time when the industry was passing through severe competitive phase and recession. Spontaneous acceptance by the cash-strapped Indian industry has gone unnoticed in the country.

There are two distinct categories of the Indian industry. The first is the organised sector, working in the environment of global competitiveness; and the other is unogranised sector serving the local consumer. Whereas quantifiable results have been documented by the established organised sector; perceptible changes leading to environmental cleanliness have also been observed in unorganised sector, though not recorded. I am tempted to refer to energy conservation efforts of the industry. Every time entries for the National Energy Conservation Awards are received, the small and unorganised sector industries are missing. They do not apply for fear of any consequential treatment from revenue authorities. The same also applies here.

Paper presented at the International Environment Conference organised by FICCI during February 2002 at New Delhi.

India's agreement on the Ministerial Declaration at the latest WTO Conference at Doha is an acceptance of the ground realities. Whereas the government was earlier pressurised by the industry not to agree to environment as an issue for task; industry's acceptance at the environmental protection route has encouraged the government to accept environmental non-tariff barriers in trade. Government's determination is truly a reflection of the amount of work done by the industry, both in the private sector as well as in the public sector. Role of FICCI in motivating industry stands out. In post-1992 era, the introduction of environmentally clean technology has also resulted when capital machinery and equipment in the industry was upgraded and modernised. That was one reason, that without specifically allocating funds for environmental cleanliness, the industry upgraded itself and this progress went unnoticed.

WTO Ministerial Declaration and Environment

The Ministerial Declaration at the WTO Conference at Doha, in November 2001, specifically mentioned the following:

> "We strongly reaffirm our commitment to the objective of sustainable development, as stated in the Preamble to the Marrakesh Agreement. We are convinced that the aims of upholding and safeguarding an open and non-discriminatory multilateral trading system, and acting for the protection of the environment and the promotion of sustainable development can and must be mutually supportive. We take note of the efforts by Members to conduct national environmental assessments of trade policies on a voluntary basis. We recognize that under WTO rules no country should be prevented from taking measures for the protection of human, animal or plant life or health, or of the environment *at the levels it considers appropriate*, subject to the requirement that they are not applied in a manner which would constitute a means of arbitrary or unjustifiable discrimination between countries where the same conditions prevail, or a disguised restriction on international trade, and are otherwise in accordance with the provisions of the WTO Agreements. We welcome the WTO's continued cooperation with UNEP and other inter-governmental environmental organisations. We encourage efforts to promote cooperation between the WTO and relevant international environmental and developmental organisations, especially

in the lead-up to the World Summit on Sustainable Development to be held in Johannesburg, South Africa, in September 2002."

The ministerial assertion on upholding and safeguarding an open and non-discriminatory multilateral trading system, and acting for the protection of the environment and the promotion of sustainable development is mutually supportive. Also a call to conduct national environmental assessment of trade policies on voluntary basis shifts the onus from multilateral inspection agencies to respective countries. How does this really work? Every country can develop its own environment assessment systems, not basing them on government's inspection and controls but further through voluntary actions by various stakeholders, of which the industry comes first. The industry will include, both the public sector as well as the private sector.

That is the status of environment in local and global consideration. The government's pledge to this process encourages the industry to operate.

Golden Peacock Environment Management Award — Showcase for Industry's Efforts

The World Environment Foundation established the Golden Peacock Environment Award in 1998. The World Environment Foundation, working in the field of conservation and development. Its main objective is to create awareness of the need and implementation of strategies for improving environmental management standards. The emphasis is on development, management and countrywide awareness of duties of individuals and corporations to maintain ecological balance by minimising adverse environmental impact, while carrying out industrial and business activities and aim toward sustainable growth. The Golden Peacock Management Award is a major initiative in attaining this goal by conferring these awards for best management of environmental practices by different organisations. The Award is recognition of the success achieved in managing environmental concerns, while at the same time, maintaining highest levels of profitability; and thus, establishing a synergy between environment and business. The Award winners are all those who have established that "environment is good business".

Categories of Awards

Awards are given in two different categories — 'Manufacturing' and

'Service' sectors. The two sub-categories are 'Large' and 'Small and Medium Enterprises' under both the categories. A unique feature of this Award is that it covers public and private undertakings, government and semi-government departments; trade and professional associations; and educational, service and research establishments including monitoring laboratories.

Impact of Awards

During the three-years (1999-2001) of presentation of these awards, it is observed that the process of application, selection and presentation has stimulated and helped organisations to rapidly accelerate the pace of sustainable, effective Environment Management System (EMS) in the organisation. It is a powerful self-assessment process and a way to build organisations 'a brand equity' on environment management. Preparation of the application for the Award helps to inspire and align the entire work force and management functions. The knowledge gained by the organisation in self-assessing their operation has often led to improved organisational performance.

Constitution of Assessment Committee and Jury

The evaluation and grading is carried out by a panel of assessors at the first stage who are people of integrity and experience in the field of environment awareness and environment audit. At the higher level the Jury committee reviews the assessment by the panel of assessors and then finally decides on the winners. During the last two years, Jury has been headed by Shri N. Vittal, the Central Vigilance Commissioner; and Lt. Gen. Surendra Nath, Chairman, Union Public Service Commission. Members of the Jury included, Secretaries to the Government of India; Chairman, Central Pollution Control Board, and CEOs of Private/Public sector undertakings.

Information Sought from Applicants

The applicant has provided information on following issues:

(i) Basic information about the organisation.
(ii) Business overview, giving an outline of the applicant's business.
(iii) Self-Appraisal Report — based on the guidelines so that

appreciation of the extent of excellence achieved by the applicant is clearly brought out.

(iv) Annual Report of the organisation—reflecting environment policy and disclosures about environment.

The Self-Appraisal Report has following elements:

(i) Corporate Governance practices focusing on environmental issues.
(ii) Legal compliance.
(iii) Training of personnel.
(iv) Procedure for Monitoring Environmental Performance.
(v) Application of Clean Technologies.
(vi) General information.

It will be observed that the guidelines are very exhaustive and assist in giving a fair response to help assessment. Very often the companies have reported that in the first year of taking the guidelines, they observed that quite a few of the activities, though carried out, were not properly documented. For lack of data they did not apply in the first year, but then opted for an application in the second year.

There were also a number of organisations who could furnish all the details as per the proforma guidelines in the very first year. Guidelines have been prepared by referring to other international awards for environment, quality and corporate governance. These guidelines are up-dated every year to incorporate the global changes in assessing the performance of organisations and industry. Following this criteria also ensures that the Indian industries follow the norms and structure as in other international industry.

Analysis of Award Winning Companies

Table 26.1: The results for the years 1999, 2000 and 2001

	1999
Manufacturing—large	
Winners	Gujarat Refinery, IOCL
	Haldia Refinery, IOCL
Runners-Up	Madras Refineries Limited
	Rajasthan Atomic Power Station
	Bharat Petroleum Corporation Ltd.
	Barauni Refinery, IOCL
	ITC Ltd., Chirala

Table 26.1 (*Contd.*)

Manufacturing—Medium & Small	
Winner	LPG Recovery Plant (GAIL), Vijaipur
Runners-Up	Salem Steel Plan (SAIL)
	TATA Metaliks Limited
	2000
Manufacturing—large	
Winners	Panipat Refinery, IOCL
	Narora Atomic Power Plant
Runners-Up	Haldia Refinery, IOCL
Manufacturing—Medium & Small	
Winner	TATA Metaliks Limited
Runners-UP	LPG Recovery Plant, Usar
	ITC Ltd., Tribeni Tissues Division
	2001
Manufacturing—large	
Winners	Chennai Petroleum Corporation Ltd.
Runners-Up	Heavy Water Plang (Manuguru)
Manufacturing – Medium	
Winner	Hindustan Coca Cola Beverages (P) Ltd.
Runners-UP	GAIL, LPG Recovery Plant, Usar
Manufacturing- Small	
Winner	Desalter Plant, ONGC Ltd., Ahmedabad
Service—Large	
Winner	ITC Hotel Maurya-Sheraton & Towers
Runners-UP	NTPC, Corporate Office, New Delhi
Service—Medium	
Winner	Rajiv Gandhi Cancer Institute & Research Centre, Delhi
Service—Small	
Winner	Koyali-Ahmedabad Products Pipeline (IOCL) Gujarat.

Significant Observations

A few significant points observed from the replies indicate the following:

(i) It will be observed that any organisation that was Runners-up in one year, showed marked improvement and could get the Winner's Award in subsequent year. Examples includes,

Atomic Power Plant and Tata Metaliks.

(ii) A number of industries also reported that they were following environment standards which were much more strict than the ISO 14,000. One classic example is of Hindustan Coca Cola Beverage Pvt. Ltd Ameenpur unit (near Hyderabad).

(iii) Every applicant was sent a feedback on "strengths" and "areas of improvement". It has been observed that in subsequent years the organisations have improved on the basis of the suggestions.

(iv) Awareness about 'environment cleaning-up' and 'environmentally clean technologies' is widespread in the industry, both in the organised sector as well as in the un-organised sector.

(v) Organised sector has opted for ISO 9000 and ISO 14000 certifications to improve their operations and also to become globally competitive. Companies have been fulfilling all the obligations and carrying out compliance of legal provisions/requirements, carrying out audits and renewing their certificates in time. Some of the smaller companies are finding the difficulty of assigning manpower to respond to all the requirements. But, that does not indicate their lack of interest in environment management system.

(vi) Larger polluting industries like petro-chemicals were the first to adopt ISO 14000 specifications, and therefore, they have well established environment management systems with proper documentation of evidence and data and facilities for training of the staff.

(vii) Organisations in sectors with public perceptions on their potential threat to environment, have also opted to document evidence to prove that they are environmentally sound. (This includes atomic power stations and heavy water plants, under the Department of Atomic Energy).

(viii) Manufacturing industry, in large as well medium and small sector, have opted for the environment management system though they are slow to start. Since their potential to release pollutants was low, these companies have attained environmental management systems more by conforming to quality production in conformity with ISO 9000. That reflects that there is a close link between ISO 9000 and ISO 14000.

(ix) Service organisations have also opted for environment management systems; the initial entrants include hotels and hospitals. Whereas hotels are totally a commercial organisation, and therefore, they have opted for ISO 14000 more strictly as compared to the hospitals that have to dispose of bio-medical wastes. Since a large number of hospitals are mostly in the government sector, the authorities have not been strict on them.

Companies Adopt ISO 9000 and ISO 14000 Simultaneously

In this regard the Golden Peacock National Quality Award for the year 2001 (presented on January 19, 2002 — just 15 days back) will be of interest. A list of Award winning companies is given in Table-2. While assessing for quality, it has been observed that the Award winning companies, whether from, the process sector or the manufacturing sector had not only opted for ISO 9000 but also for the ISO 14000. This is clear indication that there is a greater awareness amongst the industry to adopt the international standards and to improve upon that. It was observed that most of the companies who are ISO certified register only such vendors who are also ISO certified. The classic case is of the VIP Luggage Systems of Nashik, whose over 80 per cent of the vendors are ISO 9002 certified whereas the company itself is ISO 9002 and ISO 14001 certified. No doubt close linkage between ISO 9000 and ISO 14000 has been established now under the 2000 series, Indian industry's pioneering initiatives need to be fully recognised. FICCI has played a key role in creating awareness of EMS and that need be documented.

Winning Works Wonders for Industry

Winning the Golden Peacock Environment Management Award (GPEMA) has reportedly brought wide-ranging benefits for all types of business and trade, from multinationals to small family-run concerns. Some of these benefits are:

- National and local publicity strengthened customer/client base and attracted new customers.
- Enhanced staff morale and motivation — existing staff more productive.

- Commitment to environment is reinforced at corporate level.
- The Award has given a competitive edge in today's rapidly changing market place.

Conclusion

A look at the list indicate that all across the board, the industry from large, medium, small sectors are taking on the quality certification. The numbers of certification bodies and agencies and the audit organisations have increased exponentially during the last five years, indicating the scope and need for such organisations.

One can only say this has been a silent revolution about which much awareness has not been created. But means to be understood is that whereas the award lime GPEMA and GPNQA have offered an opportunity for organisations to reflect their work, it has been the initiative of the organisations like FICCI and other similar bodies, who understood the hard task during the last decade to create an awareness over the need of quality and environment system.

27

Disaster Mitigation: International Dimension

Introduction

Natural disasters have been a major concern of the global community, as the disaster can in some cases be predicted, but not avoided. Natural disasters continue to strike unabated and without notice and are perceived to be on the increase in their magnitude, complexity, frequency and economic impact. The hazards due to natural and manmade disasters has imposed threat to people, structures of economic assets and assumed disastrous proportions when they occurred in the areas of dense human habitation.

In essence disasters adversely affect the three pillars of sustainable development identified by the World Summit on Sustainable Development, concluded only on September 4, 2002. These components are: economic development, social development and environmental protection—as interdependent and mutually reinforcing pillars. Natural disasters also erode the natural resource base of social and economic development.

Modern science and technology have helped in understanding the mechanisms and computations, which result in catastrophic events of disaster and devising means by which their effects can be minimised. The advanced satellite based predictability and remote sensing techniques help in minimising the adverse impact of disaster as advance information helps in preparedness. These help in mitigatory measures though not elimination of disasters.

Paper presented at the National Seminar on Armed Forces and Disaster Management organised by the Corps of Engineers, Indian Army and the Institution of Engineers (I), Delhi State Centre on September 11, 2002 at New Delhi.

Natural Disasters — Who is Hurt the Most?

The problems of disaster are serious for the developing countries as the world's worst disasters have occurred between the Tropic of Cancer and the Tropic of Capricorn. This is also the area inhabited by the poorer countries of the world. The unfortunate part is that disasters inflict on such countries serious and repeated setbacks to progress in addition to increasing pressure on ecology and environmental conditions due to rising population and poverty. Some of the developing countries, in spite of reasonable growth during the past few decades, have remained developing only for the fact, that the result of their progress were nullified by these disasters.

Between 1960 and 1997, natural disasters have resulted in the loss of over 3 million lives and affected many more. Their economic costs are on the rise in alarming proportions; compared to 1960s, the economic loss due to disasters has increased by a factor of 8, discounting inflation.

Table 27.1: Annual Average Number of People affected by Disaster between 1987-1996: Worst affected Ten Countries of the world

Country	*Annual Average Population Affected (million)*	*Total Population (1996)* (miilion)*	*Percent Population Affected*
China	99.7	1232.08	8.04
India	56.56	944.58	5.99
Bangladesh	18.57	120.07	15.47
Ethiopia	4.02	58.24	6.90
Philippines	3.69	69.28	5.33
Australia	2.28	18.05	12.63
Thailand	1.67	58.70	2.84
Sudan	1.48	27.29	5.42
Malawi	1.44	9.84	14.63
Pakistan	1.40	139.97	1.00

*UN Department for Economic and Social Information and Policy Analysis.
Source: World Disaster Report 1998.

It is a matter of deep concern that ninety per cent of the natural disasters and ninety-five of the total disaster related deaths worldwide occur in developing countries. Table 27.1 the annual average number of people affected by disasters between 1987 to 1996. Amongst the top ten countries, China has an annual population of 99.07 million affected by disasters, and India has 56.56 million. These correspondingly represent 8.04 per cent of the total population for China and 5.99 per cent of total population for India. However, in terms of actual numbers the total population in China affected by natural disasters is more than the total of the remaining nine of the ten worst affected countries of the world. Leaving China aside, the population affected in India is almost double of the remaining eight countries. The immensity of problem for our country can be realised from this very basic fact.

Of the ten worst affected countries, only Australia is a developed country and the remaining nine are developing countries.

South Asia Regional Context

The World Summit on Sustainable Development also realised that regional context of natural acts is relevant. In regional context, South Asia, which has the largest number of poor and malnourished and is deeply affected by the terrorist activities, faces all kinds of natural disasters. Countries in this region are densely populated and are low income economies. Recurrent natural disasters have caused setback to this effort. In fact it has resulted in increasing poverty conditions in the region.

South Asian countries have diverse agro-climatic zones, each subject to particular natural disasters. Long coastal regions are prone to cyclones; arid and semi-arid regions to persistent droughts; the Himalayan mountain terrain and parts of the continental crust to earthquakes and landslides; and the near perennial rivers to periodic floods. In recent times, increased population pressures and environmental degradation have further aggravated the situation. Thus, the hazards due to natural disasters become more pronounced.

The coastal regions of India, Bangladesh and Sri Lanka are severely affected by cyclones starting in the Bay of Bengal. In the recent past, floods have become an almost annual feature of the region and the cause of heavy losses. The major rivers of the region like the Ganga, Brahmaputra and the Indus are prone to flooding — either due to heavy rains or due to fast melting of snow in the Himalayas. Floods occur with unfailing regularity in Pakistan, India, Sri Lanka, Bangladesh and Nepal, with occasional flash floods in Bhutan. As much as two-thirds of the total area of Bangladesh was affected in a single episode of floods in 1998.

Seismic disturbances are common all over the region. Latur (Maharashtra), Chamoli (Uttaranchal) and Anjar (Gujarat) earthquakes had worst devastating effect on large population of India residing in hilly, plain and coastal areas. Nepal alone has experienced 23 major earthquakes between 1890 and 1975. Pakistan has a long history of earthquakes. Bangladesh is also susceptible to an occasional seismic disturbance though of a lower intensity.

The agro-climatic diversity of the region is responsible for extremes in climatic conditions. In fact, it is not uncommon for one part of a country to be experiencing droughts while another part of the same country is reeling under the impact of floods. Two-thirds of India with an average annual rainfall of less than 1125 mm come under the arid and semi-arid regions and dry-sub humid category, which are prone to droughts. The state of Sindh in Pakistan is the country's drought-prone area, and there have been 8 major droughts in Pakistan since 1947. Sri Lanka's northern and eastern parts also suffer from droughts occasionally.

Landslides are an increasingly common occurrence in the hilly areas of the region, particularly in the lower Himalayas (Himachal Pradesh and Uttar Pradesh) as well as Kerala and the North-east and cause extensive damage to roads, bridges, human dwellings, agricultural lands etc. Environmental degradation of hilly areas has led to greater frequency of landslides.

UN Effort for Natural Disaster Reduction

Recognising the rapidly rising worldwide toll of human and economic losses due to natural disasters, the UN General Assembly in 1989, took a decision to launch a far reaching global undertaking during the nineties to save human lives and reduce the impact of natural disasters. With this aim the decade of 1990-2000 was declared as the International Decade of Natural Disaster Reduction. The programme ultimately developed an International Strategy for Disaster Reduction (ISDR).

Incidentally this was one of the international action undertaken in the preparatory phase of the Rio Earth Summit in 1992.

The objective of the International Decade for Natural Disaster Reduction (IDNDR) was to reduce, through concerted international action, especially in the developing countries, the loss of life, property damage and social and economic destruction caused by natural disasters.

By the year 2000 the IDNDR programme developed following three components.

(a) Comprehensive national assessments of risks from natural hazards, with these assessments taking into account their impact on developmental plans,
(b) Mitigation plans at national and/or local levels, involving long term prevention and preparedness and community awareness, and
(c) Ready access to global, regional, national and local warming systems and widespread dissemination of such warnings.

The IDNDR programme was evolved on the basis of the Yokohama World Conference in May, 1994, where a plan of action for disaster reduction called the Yokohama Strategy was evolved. This strategy gave guidelines for natural disaster prevention, preparedness and mitigation, shifting the focus and emphasis from '*disaster management*' to '*disaster prevention and preparedness*'.

The Yokohama World Conference, based on the assessment on the progress accomplished during the first half of the decade, formulated a strategy for disaster reduction centred on the objective of saving human lives and protecting property. The strategy called for an accelerated implementation of a Plan of Action to be based on certain variables like:

(i) development of a global culture of prevention as an essential component of an integrated approach to disaster reduction,
(ii) adoption of a policy of self-reliance in each vulnerable country and community comprising capacity building as well as allocation and efficient use of resources,
(iii) community participation in the disaster reduction process,
(iv) improved risk assessment,
(v) broader monitoring, and
(vi) communication of forecast and early warning.

During the remaining part of the decade, it called upon all the countries to unequivocally give political commitment to reduce their vulnerability through appropriate means. It also recommended that donor countries should give increased priority to disaster prevention, mitigation and preparedness in their assistance programmes and budgets.

Thus it will be seen that the IDNDR which was set up as an outcome of the preparatory phase of the Rio Earth Summit was given a concrete shape only after the Rio Earth Summit finalised Agenda-21 and defined its Implementation Plan.

In subsequent actions, the donor countries emphasised mostly on relief and rehabilitation.

World Summit on Sustainable Development — Commitment to Disaster Reduction

The world leaders expressed their commitment to better disaster preparedness measures at the World Summit on Sustainable Development, at Johannesburg.

The *Political Declaration* has in no uncertain terms emphasised on fighting disasters. Paragraph 45 states as follows:

> *"We are committed to the reduction of the economic, social, and environmental impact of natural disasters by strengthening international cooperation, deploying sound technologies, and further developing early warning systems."*

The preceding paragraph-44 can also be read as an extension of the global political commitment to fight disasters caused mainly by desertification and loss of biodiversity. This paragraph reads as follows:

> *"We agree to protect and restore the integrity of our planet's ecological system, with special emphasis on preserving biological diversity, the natural processes that sustain all life on earth, as well as addressing the process of desertification. The significant reduction in the rate of current bio-diversity loss at national and global levels is a priority to achieve sustainable livelihoods for all."*

The *Implementation Plan* includes the following:

An integrated, multi-hazard, inclusive approach to address vulnerability, risk assessment and disaster management, including prevention, mitigation, preparedness, response and recovery, is an essential element of a safer world in the 21st century. Actions are required at all levels to:

(i) Strengthen the role of the International Strategy for Disaster Reduction (ISDR) and encourage the international community to provide the necessary financial resources to its Trust Fund;
Support the Establishment of effective regional, sub-regional, and national strategies and scientific and technical institutional support for disaster management;

(ii) Strengthen the institutional capacities of countries and promote international joint observation and research, through improved surface-based monitoring and increased use of satellite data, dissemination of technical and scientific knowledge, and the provision of assistance to vulnerable countries;

(iii) Reduce the risks of flooding and drought in vulnerable countries by, *inter alia*, promoting wetland and watershed protection and restoration, improved land-use planning, improving and applying more widely techniques and methodologies for assessing the potential adverse effects of climate change on wetlands and, as appropriate, assisting countries that are particularly vulnerable to these effects;

(iv) Improve techniques and methodologies for assessing the effects of climate change and encourage the continuing assessment of these adverse effects by the Intergovernmental Panel on Climate Change;

(v) Encourage the dissemination and use of traditional and indigenous knowledge to mitigate the impact of disasters, and promote community-based disaster management planning by local authorities, including through training activities and raising public awareness;

(vi) Support the on-going voluntary contribution of, as appropriate, NGOs, the scientific community, and other partners in the management of natural disasters according to agreed, relevant guidelines;

(vii) Develop and strengthen early warning systems and

information networks in disaster management, consistent with the ISDR;

(viii) Develop and strengthen capacity at all levels to collect and disseminate scientific and technical information, including the improvement of early warning systems for predicting extreme weather events, especially El Nino/La Nina, through the provision of assistance to institutions devoted to addressing such events, including the International Centre for the Study of the El Nino phenomenon;

(ix) Promote cooperation for the prevention and mitigation of, preparedness for, response to, and recovery from major technological and other disasters with an adverse impact on the environment in order to enhance the capabilities of affected countries to cope with such situations.

The World Summit on Social Development has, for the first time, focused appropriately on Disaster Reduction Strategies and not only on Disaster Preparedness or Disaster Prevention. This reflects a pragmatic approach based on the experience gained during the International Decade for Natural Disaster Reduction. The shift in focus from 'rescue and relief' to 'preparedness and mitigation' has been appropriately understood globally and that is the reason why WSSD focused on reduction measures.

Disaster mitigation is a "collective term used to encompass all activities undertaken in anticipation of the occurrence of a potentially disastrous event, including preparedness and long term risk reduction measures". Mitigation basically involves reducing the actual or probable effects of extreme disaster on man and his environment. Mitigation is often not given the same level of priority as preparedness or recovery. This is because there is tendency to view disasters and development in terms of trade-offs with needed resources being diverted from development to disaster mitigation. Yet disasters often undermine development efforts and fritter away resources, which have been allocated to other sectors. They interrupt ongoing programmes and divert resources from their intended beneficiaries. In the long terms, it affects the cycle of sustainable development in the fragile poor economies. The global community has very appropriately focused on one component of disaster mitigation and that is disaster reduction measure.

Terrorist Activities — A New Dimension to Man Made Disasters in South Asia.

Besides the commonality of natural disasters facing South Asia, there is yet another challenge pervasive all through South Asia and can be termed as the worst form of man made disasters. That is the geo-political dimension of manmade disaster in terms of terrorist activities, and the need for fighting it. India, Sri Lanka, Bangladesh and Nepal all face the problem of terrorism in one form or the other. Unlike the natural disasters, the predictability of such manmade disaster is far more uncertain, though the vulnerability of men and property can be devastating. Whereas Nepal and Sri Lanka are facing only the internal uprising and the terrorism caused through them, India has to face the cross boundary terrorism, which is causing a much bigger strain on India's geographic and environmental security.

Conclusion

Armed Forces have repeatedly demonstrated their strength in handling the relief and rescue operations in case of natural disasters and terrorism in case of man made disasters. A closer interaction amongst the administrative machinery and the armed forces, at all time level is necessary to develop an effective strategy and action plan to ensure preparedness. As national; security is directly linked with disaster, active involvement of armed forces becomes all the more important.

28

Capacity Building Towards Disaster Management

The Disaster Management came in focus after the Gujarat earthquake in January 2001. The devastation on such a large scale attracted attention of the policy makers and the public. The Disaster Management, in varied forms, has existed even in the historic period. A major focus on disaster management through the administrative practices, initially focusing on famines necessitated the need for building a Disaster Management mechanism into the civil administration system. Though the Disaster Management dates back to 1878, when, as a result of a series of droughts in the country, the first Famine Commission was appointed. It made several suggestions, on the basis of which the famine codes were promulgated from 1883.

In the initial years, following independence, the subject of natural calamities was handled in the Union Home Ministry before a scarcity relief division was set up in the Ministry of Agriculture, to deal with the problem of food scarcity in the country in periods of drought. Subsequently, the division was upgraded to deal with other types of natural calamities.

With setting up of the Natural Disaster Management Division, in the Ministry of Agriculture, disaster preparedness and relief measures relating to food and agriculture were addressed. As the development procedures and the focus on the quality of life sharpened, different other sectors and types of disasters and crisis were identified. Since all these could not be covered under the purview of the Ministry of Agriculture alone, concerned ministries and nodal agencies have also been brought into Disaster Management procedure. With the need of

---

Validictory Address at Training Workshop on Capacity Building Towards Disaster Management organised by Indian Institute of Public Administration on November 28, 2001 at New Delhi.

the sophisticated technologies the professional institutions were also involved.

Natural disasters are not totally unpredictable, though their intensity and the impact could be unpredictable. What makes the impact of these disasters serious is the lack of infrastructure and facilities, created in the disaster prone areas, and that available ones not conforming to the disaster resistant standards. This is something easily said. Various policy makers and the administrators can be far easily blamed for inefficient practices and poor governance. At no point of time, since independence, the Central government and the State government had a booming economy, where substantial financial resources and technical manpower could be dedicated to disaster preparedness. As the total practice in the independent India started more as a relief measure in the food sector generally the practice was to wait till a tragedy occured and then provide only the relief.

Focus on various environmental concerns and disasters in the last decade of the second millennium created awareness on the need for disaster preparedness and management, to reduce the impact of natural disasters. Recognising the rapidly rising worldwide toll of man and property due to natural disasters, the UN General Assembly in 1989, took a decision to launch a far-reaching global undertaking during the 90s to save human lives and reduce the impact of natural disasters. With this aim in mind, the decade of 1990-2000 was declared as the International Year for Natural Disaster Reduction (IDNDR).

The objective of IDNDR was to reduce, through concerted international action, especially in the developing countries, the loss of life and property damage and social and economic disruption caused by natural disasters, such as earthquakes, floods, cyclones, landslides and locust infestations, drought and desertification and other calamities of natural origin.

IDNDR also provided a global mechanism for capacity building in mitigation plans, comprehensive national assessment of risks and ready access to global to local warning systems and widespread dissemination of such warning. The 1994 Yokohama Conference also suggested shifting the focus and emphasis from Disaster Management to Disaster Prevention and Preparedness. A number of other UN agencies also cooperated to ensure international cooperation and networking mechanism.

India with its large coastline, deserts and Himalayan ranges is in a peculiar situation of being subjected to natural disasters of all kinds.

Whereas we cant prevent the natural disaster, but can certainly be in a position to implement the strategies that will reduce the loss of life and property and that point was clearly and loudly made after the Gujarat earthquake. The Seattle earthquake, immediately after the Gujarat earthquake resulted in hardly any loss of human life and minimal damage to property. The intensity of earthquake at both these places was the same. Such comparative data has also alerted various scientific and engineering institutions.

The Institution of Engineers has been seriously debating on the measures that need to be undertaken to mitigate the effects of natural disasters. Under the overall guidance of eminent engineer, Dr. R.K. Bhandari, who was earlier the Adviser, in CSIR, the Institution has played an important role in relief measures after the Gujarat earthquake and developed an overall programme for micro-zonation studies in earthquake prone areas. Through its forum of Sustainable Development it has encouraged its 90 centres, spread all over the country, to look into engineering solutions connected with mitigation of disaster effects. The relief work in Gujarat was spearheaded by Shri H.P. Jamdar, the Principal Secretary, Public Works in Gujarat at that time, and the President of the Institution of Engineers. That did provide us an opportunity to study various aspects of disaster management that need to be undertaken. The Institution has looked into the issues like micro-zonation studies for earthquake prediction; structural engineering aspects for giving safer shelter and habitation. The State of Gujarat also provided a unique opportunity to study on what needs to be done when you have a combination of natural disasters like drought, flood and earthquake all at the same place. We at the Institution had worked hard in collaboration with World Federation of Engineering Organisations to identify various engineering inputs and the technological solutions for disaster preparedness in terms of safer shelters, reliable production mechanisms and relief support systems.

In all our studies it was observed that the basic concepts of Disaster Management were really not clearly understood by different stakeholders. In the case of Disaster Management, the role of government agencies in laying appropriate policies; developing linkages; establishing networking arrangements and creating capacity building institutions is extremely important so that in times of disasters the professional institutions, non-governmental organisations, community based organisations are in a position to work closely with the administrative machinery. Very often it is said, that the effectiveness of the administrative machinery has been on the decline, but the fact remains that such government agencies need to only work as facilitators, provide general guidelines and provide facilitating mechanisms to various other organisations/institutions to develop disaster management practices and in developing overall effective network in the country that can swing into action as and when required.

The Institution of Engineers (India) could certainly mobilise its members and coordinate with other professional institutions and organisations to see that a much closer arrangement for disaster management is created. However, we have always felt the need on a continuing basis to organise the capacity building programmes. The Ministry of Agriculture and IIPA have very appropriately developed this programme with the following objectives:

- To enhance the understanding of the basic concepts of Disaster Management.
- To enhance administrative capabilities to plan and implement various safety measures.
- To develop effective communication between various government and non-government organisation for timely action and disaster mitigation.

It has been observed worldwide that the technological options at times cost heavily for their development and their dissemination. But appropriate skills for management of different situations can often be developed at a low cost. A combination of best management practices, administrative acumen and technological options can provide most effective Disaster Management system in the country.

29

Mountains and the Safer Human Settlements

As 2002 is the Year of Mountains we are focusing on human settlements in mountain areas. However it is necessary to have a look at the overall mountain ecosystem — their contribution; problems; fragility of ecosystems; challenges faced by the mountain communities; environmental issues and the potential of their sustainable development.

The mountains are:

(i) Important source of water, energy and bio-diversity.
(ii) Provide resources—agriculture, forestry, minerals, recreation sites.
(iii) Immensely fragile eco-system:
   (a) Steep slopes
   (b) Vertical dimensions
   (c) Soil erosion
   (d) Landslides
   (e) Loss of habitats and bio-diversity.

Challenge to mountain people

(i) Endemic poverty:
- Harsh physical environment
- Isolation
- Lack of Communication
- Failure to transfer knowledge and technology
- Marginalisation of ethnic people

Welcome Address at the Round Table on Mountains and the Safer Human Settlements organised by the Institution of Engineers (India), Delhi State Centre on June 5, 2002 at New Delhi.

(ii) Several issues and challenges — How to integrate mountain people?

Environmental issues

(i) *Fragility of mountain systems* — delicate, steep slopes. severe climate, soil erosion, nutrient loss, landslides, limited cultivated area, over exploitation, narrow band of habitats due to altitude.
(ii) *Supporting multiple functions* — 1/5th of global landscapes, 1/10th population, capture moisture, 50 per cent fresh water supply, 90 per cent river flow in Asia, power, wood, minerals, bio-diversity, livelihood.
(iii) *Bio-diversity conservation* — new protected zones in trans-frontier—people's participation-blend of new tech. With traditional know-how — sustainable eco-system management.

Solutions/Opportunities for Sustainable Mountain Development

(i) Development of niche for strategic mountain services and products
(ii) Effective market
(iii) Environmental protection
(iv) Legislation, policies and institutional mechanism
(v) Community consultative process
(vi) Potential for decentralisation
(vii) Trans-boundary cooperation
(viii) Lessons from other regions

It is relevant to discuss the environmental issues and opportunities for sustainable development of mountain areas as well as the topic of safe human settlements.

Earthquakes in Uttaranchal during the last decade have attracted attention of policy makers to identify suitable measures for preparedness and rehabilitation. India's geographic location places it in seismic zone 4 and 5. It is, therefore, necessary to focus on preparedness and rehabilitation. Whereas the preparedness would suitably enable the humanity to minimize the adverse impact and take care of precious lives, rehabilitation package remains still the most important component of administrators. In a fragile economy like ours, the policy makers and the leaders can allocate limited scarce resources mostly for breakdown, but very little for preparedness.

The first and the most important element in preparedness as well as rehabilitation programme is designing and establishing safe human settlements. It is for this reason that at today's Round Table the Institution has focused its attention on *safe human settlements in mountain areas*. Human settlement is not only the shelter, but all the elements of activities of roads and infrastructure. Safety of structures and civil construction has become crucial. New building materials have been identified, new structural designs have been developed, and new building codes have been written to ensure that the human settlements in the mountain areas are safe. Concern has also been focused on seismo-tectonic mapping of the Himalayas and establishing techno-legal regime, and its enforcement.

Governance for safer human settlements is a concept that needs to be focused at this stage. Governance will include not only what is considered as the actions of the government either at the central level or at the state level. Governance in human settlement includes a collaborative effort in policies and actions of the government; role of funding agencies and corporate sector (the public and the private); and the role of civil society. It is here that we need to consider the three-tier governance mechanism at the central level, state level and the Panchayat level. There are a number of issues out of the 29 items mentioned in the 11th Schedule of the Constitution as responsibilities of the Panchayats which refer directly to human settlements. If Panchayats are motivated than involvement of representatives of the civil society will automatically increase.

However, it is observed that they can discharge their duties only if they are suitably assisted and supported by professional engineers. The *role of the engineering society* needs to be really understood. I have often said that engineering is art of using science and technology, traditional experience and creativity for the benefit of the humanity. The engineering profession is a major contributor to the quality of life, welfare and wealth of the people of the world. Engineering skills provide housing, civil structures and infrastructure, services for storage and distribution of water and other services vital to community health etc. Increasingly the standards of engineering knowledge and engineering practice determine the long-term welfare of people and the wealth generation capabilities of nations. In recent years it has become evident that particular engineering skills are required to reduce demand on natural resources, and to restore and safeguard the biophysical environment in which we live. The engineering code of ethics reflects strong commitment to principles of sustainable development. In that the development of the mountain areas has to be strongly projected.

It is urgent develop and promote programmes, policies and approaches that integrate environmental, economic and social components of sustainable mountain development and strengthen international cooperation, particularly at the regional and sub-regional levels through agreements, treaties and conventions.

This would include international, regional and national actions to:

(i) Implement programmes to address, where appropriate, deforestation, erosion, land degradation, loss of biodiversity, disruption of water flows and retreat of glaciers;

(ii) Develop and implement, where appropriate, policies and programmes, including public and private investments, that help eliminate inequities facing mountain communities, particularly for women;

(iii) Develop safe structural designs for buildings and houses; prepare building codes and ensure their enactment; develop appropriate building materials; conduct R&D to support traditional knowledge and skills; and conduct courses to impart skills;

(iv) Implement programmes to promote diversification in traditional mountain economies, sustainable livelihoods and small scale production systems, including better access to national and international markets, communications and transport planning, taking into account the particular sensitivity of mountains;

(v) Promote full participation and involvement of mountain communities in decisions that affect them and to integrate indigenous knowledge, heritage and values in all development initiatives;

(vi) Mobilize national and international resources with a view to promote investments, applied research and capacity building.

30

Sustainable Water Supply and Role of Engineers

Water is essential for life. It is the key resource for people's good health, for irrigating crops, for providing hydropower, for protecting ecosystems. Water resources in many countries remain fragile, more due to poor demand and supply management than due to actual water scarce. In addition to fresh water system, estuarine, near-shore and oceanic systems provide renewable food supplies, tourism opportunities, transportation highways, biotechnology supermarkets and many more benefits that are frequently overlooked or abused in many parts of the world. Pollution of water resources is on the increase in many places, and water distribution and use efficiencies are low both in irrigation and in urban water supply networks. Access to water is already a major leading factor in the socio-economic development of many countries.

There is a growing concern regarding the increasing stress on water resources, caused by population growth, unsustainable consumption patterns and uncontrolled uses.

Vision-21 on Water and Sanitation

A clean and healthy world: A world in which every person has safe and adequate water and sanitation and lives in a hygienic environment.

This Vision-21 statement was adopted by major water and sanitation agencies as the Second World Water Forum in 2000 in the Netherlands. The four key components of the Vision 21 programme for action are:

Paper presented at the Workshop on Food Security and Water Supply, organised by the Institutions of Engineers, Delhi State Centre and International Centre for Irrigation and Drainage, on October 5, 2002 at New Delhi.

- building on people's energy and creativity;
- acknowledging hygiene, water and sanitation as a human right;
- committed and compassionate leadership and good governance; and
- synergy among partners.

Critical Role of Water in WEHAB Priority Areas

Policy makers and experts at various meetings identified several key issues and challenges, with the decreasing focus on water supply and sanitation as well as the need for improved frameworks for integrated water resources management and water governance at all levels. They proposed many actions to meet the challenges — stressing the importance of taking concerted actions to use water as an entry point to achieve the goal of sustainable development.

The understanding of water's role is important within the context of WEHAB. One role as a natural resource is to stimulate sustainable economic and social development by having improved management and governance structure in place within the water sector. The second role is that of being a capitalist for sustainable development — helping others WEHAB areas to achieve their respective goals and targets.

It is important to know that these issues could also be viewed from two different perspectives:

- One dealing with the management of water related services; and
- The other dealing with the management of resource base.

Not only are there cause and effect relationships among the issues falling in the domain of different WEHAB areas but interdependencies exist among the issues within each area as well. It is necessary to identify specific issues that are most relevant to the achievement of society's goal — issued, were indicated goals and targets to be set for measuring the progress.

Water Supply and Quality Management

- More pressure on supplies;
- Fragmented management.
- Inefficient allocation systems.

Water and Energy

- Hydropower potentials;
- Renewable energy for pumping and treatment.

Water and Biodiversity

- Wetlands loss;
- Assuring water for eco-systems.

Water and Health

- Water borne diseases;
- Need for alternative sanitation systems.

Water and Agriculture

- Low water productivity and agricultural practices degrading water resources.

The three main challenges in the water sector include:

- Availability of water and its sustainable use;
- Impact on health and environment challenges.

The challenges are complex but by no means insurmountable. In responding to them a concrete plan of action — a vision, targets, implementation strategies, monitoring mechanisms and so on — need to be prepared at the country level.

Neither the challenges nor the nine key areas of action being listed below are new. The intent, however, is respond to the issues with the context of the Millennium goals and the context of WEHAB and Sustainable Development. These areas of actions are expected to have significant, wider, economic, social and environmental ramifications that would not happened by themselves:

- Providing safe drinking water supplies;
- Expanding sanitation services;
- Preparing and implementing water management action plans;

- Improving water productivity in agriculture;
- Safeguarding human health;
- Strengthening disaster preparedness planning process;
- Mobilising financial resources;
- Strengthening institutional and technical capacities; and
- Protecting aquatic ecosystems, including estuarine and marine systems.

These areas of action place water at the centre of the WEHAB framework, both as a resource and as a service provider. Making progress in these areas will help not only in achieving the water-related MDGs, but also in meeting the overarching MDG of poverty reduction, improving health conditions, exploiting hydropower potential, securing increased food production, protecting ecosystems, conserving biodiversity, and reversing unsustainable patterns of water use. In other words, addressing water-related challenges serves as a stimulus to the success of more specific actions pertaining to any WEHAB area — and making progress on WEHAB is key to the realisation of MDGs.

Most water-related problems — whether they relate to provision of services or environmental consequences resulting from poor water quality or they deal with the sustainability of systems—are usually local in nature. Possible solutions also rest in the local domain, which is why community management has become the leading approach for managing water systems. This very special characteristic of water makes it distinctly different from other WEHAB areas.

Role of Engineers

Role of Engineers in all the nine areas of action is extremely important. Realising its importance the Institution of Engineers (India) has set up the Water Management Forum in 1996, currently chaired by Mr. V.B. Patel, the former Secretary, Water Resources, Government of India. The Forum is to promote, propagate the "State-of-the-art" in respect of Water Management. The Forum puts into effect various strategies to increase the supply of available water and enable the user community to practice water saving consciousness. Gujarat has been chosen as its functional venue in view of the outstanding works done by the State in the areas of water management in irrigation and drinking water supply.

WMF undertakes publication of technical brochures and organize Seminars to disseminate existing technologies to propagate economic and cost effective water management in the context of reckoning drought over the country.

The Institution of Engineers (India) also has an international outreach through the World Federation of Engineering Organisations.

The plan of action adopted at the Johannesburg Conference refers to developing integrated water resources management and water efficiency plan by 2005, with support to developing countries, through actions at all levels to:

- Adopt prevention and protection measures to promote sustainable water use and to address water shortages.
- Develop integrated water resources management and water efficiency plans by 2002, with support to developing countries, through actions at all levels to.
- Develop and implement national/regional strategies, plans and programmes with regard to integrated river basin, watershed and groundwater management, and introduce measures to improve the efficiency of water infrastructure to reduce losses and increase recycling of water.
- Employ the full range of policy instruments, including regulation, monitoring, voluntary measures, market and information-based tools, land-use management and cost recovery of water services, without cost recovery objectives becoming a barrier to access to safe water by poor people, and adopt an integrated water basin approach.
- Improve the efficient use of water resources and promote their allocation among competing uses in a way that gives priority to the satisfaction of basic human needs and balances the requirement of preserving or restoring ecosystems and their functions, in particular in fragile environment, with human domestic, industrial and agriculture needs, including safeguarding drinking water quality.
- Develop programmes for mitigating the effects of extreme water-related events.
- Support the diffusion of technology and capacity-building for non-conventional water resources and conservation technologies, to developing countries and regions facing water scarcity conditions or subject to drought and desertification, through technical and financial support and capacity-building;

- Support, where appropriate, efforts and programmes for energy-efficient, sustainable and cost-effective desalination of seawater, water recycling and water harvesting from coastal fogs in developing countries through such measures as technological, technical and financial assistance and other modalities.
- Facilitate the establishment of public-private partnerships and other forms of partnership that give priority to the needs of the poor, within stable and transparent national regulatory frameworks provided by governments, while respecting local conditions, involving all concerned stakeholders, and monitoring the performance and improving accountability of public institutions and private companies.

Of the five challenges addressed at the WSSD was a significant challenge as follows:

> How to bring scientific knowledge to bear on decision-making and to deploy necessary resources for research and development in the areas related to sustainable development? How to ensure affordable access to critical technologies for sustainable development by poor people and developing countries?

31

Plant Maintenance and Environmental Issues

India's GDP growth over the last decade, at about 6 per cent has been a remarkable improvement over the earlier growth rate, but is not sufficient to create new employment opportunities in the country. Jobless or job reducing economic growth does not have many buyers in a country that need to provide higher employment opportunities to its people. National efforts to attract foreign/private capital into infrastructure and capital goods sector have not been successful. Such being the situation, there has been a tremendous effort to improve the efficiency of use of the stocks and capital goods already in place.

The planners and policymakers have also considered effective utilisation of capital goods as an important element in economic development of the country. A look at the mid-term appraisal of the 9th Five Year Plan and the underlying strategy of the 10th Five Year Plan Approach will make it evident.

The Mid-Term Appraisal of Ninth Five Year Plan

The mid-term review of the 9th Five Year Plan (1997-2002) indicated the following six as the major problems facing the Indian industry.

- Inadequacy of infrastructure and lack of investment,
- High cost of finance and investment,
- High cost of power, freights, port charges and congestion of ports,
- Anti-dumping duties imposed by developed countries against Indian exports,

---

Inaugural Address at National Workshop on Effective Maintenance for Optimised Plant Performance, organised by IIT, Delhi on June 12, 2001 at New Delhi.

- Non-tariff barriers imposed by developed countries, and
- The Inspector Raj on small industries.

Approach to the Tenth Five Year Plan

The Approach Paper to the 10th Plan (2002-2007) reflects that to attain a growth level of 8 per cent of GDP, it is necessary to have a growth rate of 10 per cent in the industry sector. The optimistic projection rate of 8 per cent during the 10th Plan has been fixed mainly for the reason that the scope for realising improvement in efficiency is very large, both in the public sector as well as in the private sector. However, this improvement in efficiency can only be realised if policies are adopted which ensures such improvement. The 10th plan is, therefore, expected to give high priority to identify efficiency enhancing policies both at the macro level as well as at the sector level. These policies will often involve a radical break from past practices and even institutional arrangements. Such efficiency improvements are to be gained both on the supply side as well as on the demand side. For example, for the supply side, the efficiency improvement, from the power generating plants will be in terms of higher availability of power through retrofitting the existing plants, and on the demand side it will be at the end-use level. There can be similar opportunities of improving availability of industrial plants by their effective maintenance schedules. Large investments in the stocks have to be utilised more efficiently to get higher outputs from the existing machinery and plans to achieve the national target of GDP growth.

The Approach Paper further specifies that there is sufficient evidence that there may be a considerable stock of existing capital assets, which are either lying idle or have never been used to their full potential. Bringing such assets into full productive use can certainly reduce the resource requirement by. The second area of idle capacity stock is in the public sector enterprises, where financial and managerial problems have prevented adequate investment in balancing and/or upgradation leading to inadequate utilisation of the capital assets or to the lack of competitiveness. Prolonged period of insufficient maintenance expenditure also have similar effect. Mere infusion of fresh funds is unlikely to solve this problem since the roots of the malice are deeper, and usually lie in the domain of policies and excessive interference in the management of these enterprises. Measures to unlock such potential capacities through institutional change can give rich dividends in the country.

The Approach Paper has also emphasised on sustainable development, that is, environment-friendly development. This environment-friendly thrust has also focused on efficiency of operation and indirectly better use of existing infrastructure. Conservation, protection and preservation of the environment are being emphasised which directly or indirectly are related to efficient use of capital 'stocks' and 'raw materials'. At the plant level its performance links environmental issues with effective maintenance. I will, therefore, like to deal with the environmental issues and manpower development for producing quality in maintenance.

I will therefore like to speak on environmental concerns that are linked to the optimum performance of the plant and machinery.

Industry and Sustainable Use of Resources

The pace of development and consumerism by the galloping population growth has exerted enormous pressure on our natural environment. The stress on natural resources is reflected in the form of polluted air, contaminated water, diminishing forests and extension of numerous species of flora and fauna. While managing resource sustainability, it is also necessary to care for those who depend on these resources for their livelihoods. Increasing production of goods without addressing sustainability of the resources, on which production is based, will sooner or later run into declining productivity. This leads to the issue of rehabilitation of degraded resources and adoption of policy measures to promote sustainable use of resources for basic human needs. As the industrial plants and machinery are crucial in conversion of such natural resources into products of use, it is relevant to focus on efficient capital goods and their effective maintenance to ensure their sustained availability at highest possible efficiency of operation. On the other hand the sustainability of natural resources for raw materials is directly linked to effective maintenance of plants and machinery as the more effective the conversion the better the conservation of resources. A reflection of linkage of financial benefits out of effective maintenance with their positive impact on environment has already enthused the corporate management to support such an approach. Also the management of major industrial houses are alert to environmental issues.

In a recent case of assessment of awards for the industry for their environmental performance, it was reported that while working on hot water lines workers use asbestos based hand gloves. No doubt that is the practice in major companies round the country. On observing such an action, the assessment team enquired from the CEO of the company, if they had made any effort to find an alternative to use of asbestos based products. He was also explained that although it is legally right in India to use such gloves, but the use of asbestos based products have been banned in most of the developed countries. The response was prompt, and the CEO said, that so far it was not done. But they will try to find the alternative immediately. This conversation took hardly three minutes on the telephone and spontaneous response of the CEO is indicative of the fact that the management of the company was committed and alert to introduce latest environment friendly working conditions.

This is not an isolated example. My interaction with a number of companies indicates that larger number of corporates are open to new ideas and quality concepts that could help them attain quality in end product as well as in operation; and the environment management. ISO 9000 and ISO 14000 certifications have now become the order of the day rather than isolated incidents. The process for such certification is cumbersome and many a time the smaller companies with limited manpower after taking the certificate do not renew as they are not able to fulfil all the formalities. But, on the other hand, larger companies with technically qualified manpower devote sufficient time on following the certification norms and improving on them. There are a number of cases where they are now following standards and norms, that are much more rigid than those laid down in ISO 9000 and ISO 14000. This does not happen out of the blue. The larger companies are able to allocate substantially high funding for training and re-training of their manpower. The corporate decisions are able to reap benefits on a large scale.

My visits to a number of industries, varying from petroleum sector refineries to power plants, to staple yarn manufacturing units, to IT based companies indicate that most of the leading edge companies in India appreciate following the international norms and standards for quality and environment because there is a board level commitment. Such an approach has helped them to look for quality at every stage, and obviously on higher availability of plant and machinery through effective maintenance. This is included in the good corporate governance practice. The effect of board level commitment is that the quality concept percolates down to every level and every activity of the plant.

Environment Preserving Strategies and Business Opportunities

Although Dow Chemicals was the first company in USA to have put their heart into supporting environmental management practices and reap the benefit; the case of Shell Petroleum in U.K. has been discussed publicly recently for the benefits they have derived by focusing on environment management. They have found that the environment management practices have given them new fillip and larger business opportunity and social acceptability. Even in India, the case of industries like Indian Oil Corporation (mainly their Refineries Division); Reliance Petro Chemicals; Grasim Industries; ACC and Vikram Cement and Air Tel are indicative of the fact that quality and environment management aspects help these companies to focus on better plant availability and better product on a sustainable basis. A focus on environment preserving strategies are now considered as business enhancing opportunity.

The synergy between industry and environment is now considered an important element in organising sustainability of the whole ecological system. This year has special significance as this is the last opportunity for us to define and develop inputs that can be presented at the World Summit on Environment and Sustainable Development at Johannesberg in June 2002. It has not been only for 30 years (since the 1972, Stockholm Conference) that efforts were made to protect natural environment. Efforts were initially made a century before the Stockholm meet to protect the environment from industrial pollution. The effort to save the natural environment date at least to the 1870s, when Switzerland tried to establish a regional agreement to protect the nesting sites of migratory birds. But the move to internationalize environmental policymaking did not gain serious momentum until the 1970s. Up to the time of UN Conference on Environment Development in 1992, at Rio, governments had adopted more than 170 environmental treaties on subjects of shared concern; acid rain, ocean pollution, endangered species protection, hazardous wastes export, and the preservation of Antarctica, as well as ozone depletion. More than two-third of these agreements had been reached since the landmark U.N. Conference on the Human Environment was held in Stockholm in 1972. Governments have also deployed international institutions such as the United Nations Environment Programme (UNEP), created at the

Stockholm conference, to negotiate treaties and help make them work. Lets wait and see as to how many more treaties come in effect by the time the UN World Summit an Sustainable Development takes place in June 2002 at Johannesburg.

The environmental issues are intricate and interwoven. Wind currents, rain pattern and river streams carry pollutants hundreds or even thousands of kilometres from their sources. Some 200 river basins are shared by two or more countries. On an even larger scale, the global environmental problems of ozone depletion, climate change, deforestation, and the loss of the earth's biological diversity threaten all nations. Anyone country acting alone is powerless in the face of these problems. Some of the global initiatives in terms of Protocol to control emissions have not been ratified in time, and the developed countries are not willing to fulfil their commitments.

Lack of International Approval of Kyoto Protocol

In spite of all the international efforts the global response has not been very encouraging to control activities that can lead to climate change.

In the first three months of this year, the Inter-governmental Panel on Climate Change, which is jointly sponsored by the UN Environment Programme and the World Meteorological Organisation (WMO), realised its assessment based on the available data which present a chilling picture of a warming world. It warns of adverse consequences such as the melting of glaciers and polar ice caps, leading to rising sea levels. It predicts more extreme droughts, floods and storms, and significant changes in the functioning of the critical ecological systems such as forests and coral reefs. Warmer and wetter conditions would increase the spread of infectious diseases such as malaria and yellow fever. And the inundation of low-lying islands and coastal areas could lead to the displacement of hundreds of millions of people.

In this light, it looks outright irresponsible that the Sixth Conference of the Parties to the United Nations Framework Convention on Climate Change failed to reach agreement last November in the Hague on how to finalise guidelines for reducing greenhouse gas emissions. They were unable to agree on the implementation of the Convention's 1997 Kyoto Protocol, which contains binding targets for reducing greenhouse gas emissions. One more attempt will be made in Bonn, Germany, next month and we should all hope that the State Parties this time should live up to their obligation, undertaken in the Millennium Declaration, to create conditions for the entry into force of the Kyoto Protocol by the year 2002. If this does not happen, the global warming will continue unhindered, and in this respect, the recent decision of the UN Administration to turn away from the Kyoto Protocol looms ominous, because, as the UN Secretary-General has said, developed countries, particularly the United States, which produces the most greenhouse gases, much take the helm in stemming global climate change.

There are some good news as well. Public awareness has grown. Civil society is engaged. The private sector is beginning to seize the opportunities of green technology. Partnerships are being formed. The global machinery of policy-making and governance is moving in the right direction. The world has the human and material resources to place its economies on more sustainable footing. On 23 May, in Stockholm, the Convention on Persistent Organic Pollutants, banning the world's most dangerous pesticides, was adopted and opened for signature. And, in a historic move, the United Nations Commission on Human Rights has recently concluded that everyone has the right to live in a world free from toxic pollution and environmental degradation, a decision which was hailed by the UN Environment Programme.

It is most appropriate to look at some of the observations on the impact of man made activities on environment and the opportunities that environmental activities offer to the industry.

Environmentally Sound Technologies (ESTs)

It is difficult to make a specific assessment of the environmental damage caused due to developmental activities. However UNDP tried in its Human Development Report 1998 to reflect environmental damage cost to natural resources in a few selected countries. Various estimates are only indicative. These highlight the possible dimension of the problem, but does not exactly quantify the problem.

While the situation is grim, it does provide the opportunity to introduce environmentally sound technology (EST) solutions on a massive scale. The enormity of the damage caused is reflected in the figures for some of the countries, which also give an idea about the efforts needed to clear. For example that for China it has been estimated that an investment of US $ 50 billion is needed in technologies to mitigate greenhouse gas and acid rain emissions. Such investment is already beginning to happen. All these investments are being made under the GEF or through the commercial routes. In every case the investments are being routed through the industry alone. China has invested (in 1990s) at a rate of about US $ 2 billion per year for clean technologies.

Table 31.1: Estimated Cost of Environment Degradation in Selected Asian Countries

Country	*Year of Period*	*Environmental Damage*	*Annual cost (US$ billion)*	*Cost as % of GDP (per cent)*
1. China	1990	*Productivity losses caused by soil erosion, deforestation and land degradation, water shortage and destruction of wetlands. *Health and productivity losses caused by environmental pollution in cities.	13.9-26.6	3.8-7.3
2. Indonesia	1989	*Health effects of particulate and lead levels above WHO standards in Jakarta.	6.3-9.3	1.7-2.5
3. Pakistan	Early 1990s	*Health impacts of air and water pollution and productivity losses from deforestation and soil erosion.	1.7	3.3
4. Philippines	Early 1990s	*Health and productivity losses from air and water pollution in the vicinity of Manila.	0.3-0.4	0.8-1.0
5. Thailand	1989	*Health effects of particulate and lead levels above WHO standards.	1.6	2.0
6. India	1992	*Costs of urban air pollution, water pollution, soil erosion, land degradation and deforestation.	10-13.8	4.5-5.0

Source: Human Development Report, 1998.

Situation in India is no different. With the economic reforms and impact of globalisation the industry has modernised very fast till the recession set in later half of 90s. They have realised the need for clean and cleaner technologies for production.

Industry, mainly in the industrialised countries, is using ESTs on a wider scale. The global market for environmental products such as water and wastewater treatment, waste management and recycling, air pollution control, noise reduction systems and services, is growing fast. It has increased by around 5 per cent every year during 90s. Certain estimates indicate that total market for ESTs was US$ 300 billion in 2000. The United Nations Development Programme (UNDP) projected the global compliance market (including ESTs required to meet mandatory environmental standards) to reach US$ 500 billion a year by 2000. It forecasts that the future global market for environmental investments will expand to US$ 300-600 billion annually for pollution control, a total of US$1 trillion for power from 1993 to 2000, and a total of US$ 250 billion for energy conservation over the next 20 years, upto year 2020.

If this is the dimension of investment in ESTs, then effective availability of plants through quality maintenance will be crucial.

The market for ESTs has grown for two main reasons:

- Companies have been required to meet increasingly strict regulatory rules, national regulations and international treaties for pollution and waste reduction. For many companies, the first step has been to use end-of-pipe technologies, which is a short-term and expensive solution. The ultimate goal should be to move beyond this and use cleaner production processes.
- Many corporate leaders, accepting that sustainable development is an integral part of the business agenda, have moved proactively to improve their companies' environmental performance without waiting to be mandated to do so. More and more companies have recognised that minimising, better still eliminating, undesirable harmful effluents, emissions and wastes, improves their performance both environmentally and economically. An emphasis on technological innovation has been a feature of this approach.

There is abundant evidence of the considerable financial benefits from using ESTs. The return on the investment has been significant and, as the World Business Council for Sustainable Development says, companies that introduce new technology to improve environmental performance before mandated to do so by regulations cut costs and boost competitiveness.

It is agreed that many profitable investments in new, cleaner technologies are not being made because enterprises and financial institutions do not recognize the cost reductions and revenue increases resulting from greater efficiency of resource use, reduced waste and pollution, and increased sales in environment-sensitive markets. Enterprises and financial institutions need training in the use of investment appraisal and project formulation that identify the financial benefits to enterprises of cleaner production.

Smaller companies, deprived from an essential driver and benchmarks and lacking the capacity or the courage to take action on their own, are left adrift with the static mindset, that environment management are expendable "cost-only" efforts good only to add red tape to their bottom lines.

In the face of mounting business globalisation, companies cannot but expect to stay competitive by focusing alone on traditional areas of comparative advantages such as capital, labour, energy and raw material. In fact today's globalisation trend is increasingly making the very notion of comparative advantage obsolete.

Adequately enforced market based environmental regulatory regimes may help achieve just that at least cost. It provides a framework needed by operators to make better informed and economically sensible decisions in the face of increasingly complex and interrelated economic and environmental issues.

Conclusion

Any company can now source low cost inputs nearly anywhere and new rapidly evolving technologies can help offset most of the disadvantages of costly input. To stay in business today, it is not sufficient to have resources. What makes a company competitive is how productively are these resources allocated and applied in the value chain, how it differentiates itself from competitors and how well it anticipates demand.

Effective maintenance for optimised plant operation is an important element of this value chain and can give companies lead over its competitors, in an environmentally sustainable manner.

32

Environment Protection and Maintenance in Industries

Introduction

Recent advancement under the economic reforms regime indicates that foreign direct investment and the private investment into the industrial sector have not been on the level that will boost employment opportunities through industrialisation. The Reserve Bank of India has fixed the GDP growth in the year 2002-03 at 5 per cent downgrading it from 6.5 per cent. This directly reflects the impact of recent recessionary trends in the market. Surprising enough such a regime is taking place with the interest rates have also been falling in the market. The Tenth Five Year Plan, approved by the Cabinet on the same day hopes for a GDP growth of 8 per cent during the next five-year period. It is also expectated that Rs.78,000 crores will be raised through the disinvestment route. The projection is also that our FDIs will rise to US$ 7.5 billion per year during the Tenth Plan period.

The simple puzzle one has to solve at this stage is to study the strategy that the government has outlined that will allow it to move from 5 to 5.5 per cent GDP growth in the first year of the Tenth Five Year Plan to overall increase at 8 per cent. One would really wonder as to how that will be achieved. The Plan no doubt indicates that 50 million jobs would be created through the agriculture route but that certainly does not mean that the industry can survive in an environment of jobless development. If the capital investments are difficult to be achieved, then the other possibility is to improve performance

---

Paper presented at the Workshop on Industrial Maintenance for Economic Growth, organised by the Indian Institute of Plant Engineers, during November 29-30, at New Delhi.

parameters for the existing plant, machinery and equipment and also to improve their longevity through upgradation and retrofits.

The power sector has amply demonstrated that through such measures, it had been possible to substantially improve power generation capacity from the outdated power plants that has helped us to tide over the situation in the Eighth Five Year Plan as well as in the Ninth Five Year Plan.

The Plan of Action of the, World Summit on Sustainable Development, popularly called the Johannesburg Conference has emphasised on development of a 10-year framework of programme in support of regional and national initiatives to accelerate the shift towards sustainable consumption and production to promote social and economic development within the carrying capacity of ecosystem by addressing and, where appropriate, delinking economic growth and environmental degradation through improving efficiency and sustainability in the use of resources and production processes and reducing resource degradation, pollution and waste.

The plan also focused on increasing investment in cleaner production and eco-efficiency in all countries through incentives and support schemes and policies directed at establishing appropriate, financial and legal frameworks. This would include actions at all levels by taking steps that will:

(i) Establish and support cleaner production programmes and centres and more efficient production methods by providing, *inter alia*, incentives and capacity-building to assist enterprises, especially small and medium-sized enterprises and particularly in developing countries, in improving productivity and sustainable development.

(ii) Provide incentives for investment in cleaner production and eco-efficiency in all countries, such as state-financed loans, venture capital, technical assistance and training programmes for small and medium-sized companies while avoiding trade-distorting measures inconsistent with WTO rules.

(iii) Collect and disseminate information on cost-effective examples in cleaner production, eco-efficiency and environmental management, and promote the exchange of best practices and know-how on environmentally sound technologies between public and private institutions.

(iv) Provide training programmes to small and medium-sized enterprises on the use of information and communication technologies.

Environmental Concerns in the Industry — Status

Environmental concerns among a small group of large companies is growing in number. For most companies dealing with the environment is still primarily about compliance — a costly, cumbersome and unwelcome process. Environmental concerns, however, shape every aspect of strategy and action.

The chemical industry was the first to be caught in the controversy of environmental pollution in the 50s. The instantaneous response was to deny everything. But there was no reliable data. Government and the private industry were projecting their data to substantiate the respective argument. It took a long time for the industry and the government to establish a dialogue. Once the dialogue took place the chemical industry was quick to respond, which was beyond 'compliance alone' and addressed the global environmental issues on 'pro-active' to the 'voluntary' basis and this obviously meant emphasis on delivery of action, required to protect the environment. Subsequent global industrial activities reflected that responsible care and collective action clearly makes sense for a lot of reasons including, sharing the burden, the risk of being harmed by others' actions, and so on.

The industry also realised that courting long term prosperity through green marketing, or through claims that are not sustainable, is asking for trouble. The industry has to set its priorities within the environmental arena. It is really a struggle because suddenly the environmental compliance is thrust by the government for which the industry is not prepared.

Major industries have also indicated that they are perfectly capable of preparing model laws and regulations, usually on a national basis, in response to a threat, but that is too late. These companies have seriously to think if they can put them together in advance of the public outcry, so that they are seen as the solution rather then the problem. The industry needs to work with governments to get pollution prevention installed in manufacturing, instead of resorting to end-of-the-pipe treatment. The industries can then make money on it. This is what the Business Council for Sustainable Development called 'eco-efficiency' — 'economic and environmental efficiency.'

When an assessment is made, and corporate decisions are taken on capital expenditures, it becomes necessary to identify the additional capital that has to be spent on a plant to achieve the degree of environmental protection being aimed for, and then assess whether that piece of capital is going to add or subtract from shareholder value in the long term. Some major companies have, in the past, abandoned projects after doing this analysis. They have chosen not to build a plant at all because they could not afford to put in as much environmental protection as world wanted.

Another important input for corporate planning is to develop a body of knowledge that gives a fair valuation to key raw materials and energy. If someone knows, as to what it cost to grow a tree or to process a barrel of oil, to generate a calorie or a BTU or whatever, then he can build a bridge between pollution reduction and pollution prevention. It is an issue that has been widely discussed and that takes to full cost accounting. The industry has realised that the problem with current accounting is that people do not allocate environment cost to their products.

Environment Issues in Corporate Decision-Making

The synergy between business and environment calls for a self-supporting market mechanism working independent of government's props like subsidy. The synergy cannot be effective unless appropriate elements are built into the environment policy and the economic calculus of industry and business on a reasonable time span basis of sustainability. Budget speech of the Finance Minister of India for 1998-99 also reflected the need for environment friendly business growth. Though the NGOs at WTO meet at Seattle in November-December 1999 opposed non-tariff barriers based on environmental and human rights issues, they never opposed environment-friendly business growth.

The Driving Forces for Environment Protection

India is peculiar in that it has the characteristics of the developed economies as well as the developing economies. Therefore, the driving forces include the following elements:

- Pressure from central government, including increasingly stringent regulations.

- Improved enforcement of environmental regulations by newly empowered and resources policing agencies.
- Public pressure.
- The corporate environmental policies and programmes of the larger international companies who are forming new alliances.
- International donor assisted initiatives in environmental protection.
- The emergence of ISO 14001.

Certain Basic Queries Leading to Corporate Issues

The corporate strategy may have to critically analyse some of the inputs by developing answers to the following queries:

- What role does the corporate strategy play in finding environmental solutions?
- Within the environmental arena how are the priorities decided?
- What kind of a linkage has been developed between the industry and the government to respond to environmental regulation compliance?
- Can Market Based Instruments alone take care of the environmental concerns?
- How does the industry define environmental cost? What role does waste reduction programme play in environmental cost?
- How does the industry audit total spending on the environment?
- How are managers evaluated on the environment?
- How much time does the Chief Executive spend on environmental issues?

Conclusion

Environment and Business have now established a synergy that is helpful to both. Good environment practices offer better opportunities for business; and in turn good and responsible business houses adopt environmental measures both as business option as well as social and environmental responsibility.

All the existing industries are adopting environment protecting devices and gadgets during their maintenance schedules, by adopting retrofits or practices. It makes economic sense to link scheduled predictive, preventive and break down maintenance with environment protection.

33

Improving Public Acceptability of Hydro Projects

Introduction

Hydro resources have their immense potential for growth of developing countries, mainly in the South and South-east Asia. Such a development involves large dams as an inseparable unit, and that has been an issue of public debate. The debate has focused on potential risks of large dams on the civic society. Risks projected due to the failure of large dams are not substantiated with data. Natural disasters, like earthquakes and volcanoes are observed to be two major disasters, that affect human life on a large scale. Societies have continued to live in volcano-prone areas as well as earthquake-prone areas. The coastal belt along San Francisco, USA and the whole of Japan are the areas marked as extremely seismic zones. Even after the severe earthquake in the city of Kobe, Japan has continued to build sky-scrappers and is now in the process of building the tallest building in the world, at Nagoya Station.

Human ingenuity has found solutions to various possible dangers and risks due to natural calamities. Possibility of dam burst has to be weighed against the technological developments. Since 1929, and even before that, as part of the World Power Conference, knowledge about the design, construction, maintenance and operation of dams, not only large dams, but all dams has been developed and spread by the interchange of expertise at various professional congresses and committee meetings. In this way, the structure have gradually become safer and more cost effective.

Paper presented at the ICID Silver Jubilee Lecture Series on December 24, 1999 at New Delhi.

Barriers due to Public Perceptions of Large Dams

Large projects have to face certain barriers. These barriers include the following:

(i) Adverse impact on bio-diversity due to construction of large dams.
(ii) High costs of hydro-projects.
(iii) Long gestation periods of the projects.
(iv) Risks due to bursting of dams.
(v) Resettlement and rehabilitation of oustees from the project site.

The object of my paper is not to highlight the issues of technical viability or financial success of major hydro projects but to deliberate on issues that have generated (necessary or unnecessary) discussion on the global level and have often caused extreme delays and cost over-runs in hydro projects. I will now like to deal with each one of these issues in detail.

Adverse Impact on Bio-diversity During Construction of Large Dams

Environmental issues are globally covered under two major conventions:

(i) The Bio-diversity Convention, and
(ii) The Climate Change Convention.

Whereas the *bio-diversity convention* focuses on global and local actions on flora and fauna status in different geographic locations, the *climate change convention* mainly considers local actions that have global impacts. Large hydro projects do affect, to a certain extent, the flora and fauna and thereby the issues are covered under the bio-diversity convention. Every development project will have some adverse impact on the bio-diversity and that is just unavoidable. At the same time, the hydro projects, mainly the hydro power projects have substantial benefit in terms of savings in emissions that would have been caused by use of fossil fuels in thermal power generation plants. On a global basis the WEC projected in 1996 that these hydro power plants result in saving about 13 per cent of global emissions that would have been caused if thermal power generation route was adopted.

For those who wish to consider purely from the environment angle the impact of hydro projects has to be viewed, both for its bio-diversity impact as well as the climate change impact. No specific studies have been made to counter the negative impact of one by the positive impact of the other and as such it would be premature to quantify such benefits. If these benefits cannot be quantified they can atleast be indicated. But one has really to see the localised impacts and weigh them against the global climate change impact.

The opponents of large dams or projects sometimes perceive that runoff in many areas in the world is too erratic to be useful; and they consequently advocate many small dams and schemes, rather than a single big one. They do so in the mistaken belief that small dams caused less social and environment damage than a large one. This is patently false, as can be easily proven. It goes out from the premise that damage is roughly proportional to area flooded. Mr. Theo P.C. van Robboroeck in his paper titled 'Benefits and Concerns about Dams' states, "Geometry tells us that a small object has a greater surface area in proportion to its volume than a large object. The difference may be significant: for instance, when the volume of a cube doubles, its surface area increases by only 60 per cent. This implies that, all other things being equal, the land area inundated by 50 projects with a capacity of 20MW each would be greater than the area inundated by a single 1000MW project, although the total capacity is identical. In my 1996 Geoffrey Binnie Lecture to the British Dam Society, I illustrated this phenomenon with two real case examples. In Britain, the total volume of the 327 smallest reservoirs in the British of Large Dams would be needed to replace the volume of the largest, called Quoich. Their total submerged area would be 3.5 times Quoichi's! Similarly, in South Africa, the 422 smallest reservoirs in our Register would be needed to replace Gariep's volume. In that case the aggregate area would be 2.22 times larger. An economic comparison would even be more negative, in view of the loss of advantage of scale. Which is very pronounced for such projects. Consider in this regard for example that 422 separate spillways would be needed!"

High Cost of Hydro Projects

Role of water as a key infrastructure ingredient has remained virtually unexplored. Development of hydro projects, for example in India, the River Valley Project has become a life line for progress and prosperity. A fear syndrome has been created in the past two decades against river valley projects by ignoring their need and tremendous benefits. Lack of electricity and water has resulted in blocking many potential economic development activities.

As most of the economically viable hydro sites have been explored, the available sites are in difficult geographic locations, in fragile environment and needing much more effort for executing the project, and therefore, high cost of the project and longer gestation periods that have to counter many unforeseen situations and cost overruns. Whenever such projects are viewed as a single application project, for example, for hydropower development, the project costs are always high.

Experts have indicated that a basic plan for irrigation and flood control, which is evaluated on the basis of benefit/cost ratio, would have two purposes but only one objective and that is national economic development. While such distinctions are necessary to obtain benefit cost ratios, multi-objective water resource planning includes the necessity to improve environmental quality through water management. Multi objective planning, in other words, broaden the application area and indicates to the planners that they need not evaluate objectives solely on the basis of monetary benefits and costs. It is therefore necessary to look into the hydro project development as a multi-objective and multi-purpose project. In that case, the investments can be shared into different sectors and also into different projects.

Thus to justify the high cost of hydro projects, it is extremely necessary that different components of project costs are attributed to different purposes and different objectives. As a result the financial and social cost benefit can be in favour of hydro projects. At the same time, while making such projections the environmental benefits should be fully internalised.

Difficult sites will incur longer gestation periods, but the project delays need to be avoided/curtailed. It may be advisable to plan projects without any short-cuts and building multi-objective/multi-purpose modules that can show results at shorter intervals during the project period.

Examples can be building schools, health centres, training centres for income generation activities, laying out facilities for resettlement of potential oustees etc.

Resettlement and Rehabilitation of Oustees from the Project Site and Role of Stakeholders and NGOs

Resettlement and rehabilitation of oustees from the project site has been taken up by the stakeholders as well as the NGOs as a major issue needing appropriate solution before taking up a hydro project. This action attracts attention of activists in all major development projects. There is no ideal solution for resettlement and rehabilitation. It is relevant to observe that Mangalore Power Company, the local company of Cogenetrix, a US based power generation company, undertook a major public relations gamble to win over the local population in the project area in Dakshin Kannada district. According to reports not even a week ever passed without the company sponsoring some local event or the other. The company went out of its way to contribute its might to religious institutions notably to build temples and to improve the lives of fisherfolks' children. The company sponsored virtually every sport from local cricket to the famous buffalo race held in the slushy fields. The Mangalore Power Company had already started work on a Rs.12 crore rehabilitation and reconstruction project to assist those to be displaced by the project. But this exercise proved futile and did not stop the Dakshin Kannada environment activists from opposing setting up the 1000MW-power project at the site. Cogenetrix made a serious mistake in timing — they did all this even before the project started — off. Such a pro-active move alerted the environment lobby and raised doubts about dealings of the company. Similar action after starting the project could have been extremely helpful. All major companies have 'public awareness' funds that are being used for 'public relations'. PR can help but not substitute for solutions on hard realities.

Sardar Sarovar Dam project site is the worst affected by delays in the history of large dams in India. Several state governments have come up with their optional resettlement and rehabilitation plans and have, in vain, tried to convince the project-affected stakeholders that the best possible solution is being provided to them. The NGO activists have opposed the move tooth and nail and have not allowed work to progress at a speed, one would desire to.

It is time to look back and carry out introspection to understand as to what has gone wrong. The Government of India has been consulting professionals, experts and representatives of state governments for their opinion. This opinion has been discussed in inter-ministerial official committees. All such inputs possibly encourage government to conclude that the decision is based on sound fundamentals of technical, financial and administrative practices. The government finally assumed that the stakeholders will accept the decision of the government. This worked to a great extent, till the late 80s. The 90s witnessed the era of liberalisation and privatisation of economy and greater role for NGOs in public projects and public policy making. The NGO movement in India is becoming stronger by the day.

The NGO acceptance was witnessed on the occasion of the WTO Conference, at Seattle, USA where NGOs stalled inauguration by five hours and forced the President of USA to agree for larger participation of NGOs in the WTO activities. This situation would not have been witnessed in late 80s, when NGOs were considered anti-establishment, and therefore, anti-development or anti-growth. The situation has substantially changed.

The NGO movement has been a phenomena of the post-World War–II period during which against the background of atrocities towards the Jews, NGOs took it upon themselves to monitor any irregular behaviour by one group of people against another, across the world. They became important representatives of civil society and became conscience keepers of governments throughout the world. Many have been involved in Human Rights and Development. This basic characteristic still continues. The third world NGOs are different from the industrial country NGOs in their agenda. The action at WTO indicates that there is similarity of views and similarity of approach amongst NGOs from the developed as well as the developing world. The Seattle incident has brought focus on NGOs, who asked for better transparency in action. Their argument was accepted that the economic issues cannot be debated and decided mainly by governments and corporations at the inter-governmental meetings. Global networks of NGOs have emerged as a new force that is multinational in composition and globalised in vision. Ever since the Earth Summit 1992, NGOs have been called upon to play major role. By focusing on the environment, they gained a reputation of being sensitive and missionary. The NGO movement rapidly spread to other sectors like social development, women's development, food and nutrition security etc.; the issues that were discussed at the UN Conferences during the decade of the 90s.

The present day NGOs are a multi-disciplinary and highly professional groups of people who are dedicated to a certain cause.

This is not to say that those who hold the strings of the purse that funds them will not like to influence the decision making of the NGOs. Such movements of NGOs may be led by a scientist, an economist, a historian, an author or an activist. The group always consists of professionals from other disciplines that are relevant for their work. Therefore, while judging the characteristics of a movement, one should not lose sight of the strength of that movement by considering only the leader. Every NGO has invariably tried to have a front man, whose image makes news and attracts not only the public but also the power that be. If land mines could be stopped, that was possible because the NGOs got the support of Ms. Diana.

The thrust of the argument is that with the changing times where NGO movements are becoming stronger and gaining support at local, national and international levels through networking and their clout is taking them to policy making process, it is extremely necessary that a serious and sincere dialogue is conducted with the NGOs. Perceptions can differ and optional approaches can be found to identify acceptability of common view point. As mentioned earlier, the multi-disciplinary workforce of NGO groups is as well equipped to argue as the government. There can be two different paths. The first is to think that the NGO movement is basically 'obstructionist,' and not make an effort to involve them in development. The other approach is to involve them into project development, design and implementation strategies. Their perception of projects can differ. These perceptions may be existing because of a gap between government's projections and NGOs' understanding of the project objectives. Sincere effort is required to bridge this kind of gap as far as possible.

There are two distinct categories of NGOs. Those with activist approach have wider visibility, and consider the other category of NGOs involved in national development work, under government funded projects, as the government controlled NGOs. This second category of NGOs, who can be termed as development NGOs, are accountable to the government for the public funds utilised by them. That does not mean they toe the government's line of thinking or will be forced to do so. They have done yeoman's service by assisting government in low cost, dedicated execution of national projects; and providing professional inputs for decision making.

Certain NGOs have realised the need for large hydro projects, and therefore supported major projects. Prominent amongst them is Shetkari Sangh. Since their number is low and they do not carry the clout like the activists, their support does not create much public awareness.

This is to illustrate that the major issue is not government's approach versus the NGO or stakeholder approach. Amongst the NGOs there is a group supporting large hydro projects, and the other opposing the same. This observation leads us to believe that a continuous dialogue with the NGOs can bring about a better understanding.

My assertion is that there is no short-cut method of building rapport with NGOs or the project oustees. One cannot buy them over, nor one can win them over just by throwing baits at them — as was done by Cogenetrix. It is extremely necessary to give larger share in policy making to the stakeholders and NGOs by involving them from the stage of perception and finalising the strategy for resettlement and rehabilitation which will perceptibly be seen as a programme leading to better quality of life and better income generation possibilities.

New approach will have to be followed and government will have to take benefit of new movement of privatisation, liberalisation and globalisation.

One can always argue that this kind of an approach can be discussed for projects that are still to be started. But, then what about ongoing major projects, like Sardar Sarovar Dam and Tehri Dam Hydro projects. There have been differing perceptions that have caused huge losses to the national exchequer and has also deprived the common man of possible benefits from such schemes. It is high time, that a new look or a new approach is followed for finding a suitable solution. In the past whenever a suggestion was made for providing a new look, the ball was thrown back into the court of the same persons into official positions. It is necessary to throw open issues for public debate. It is necessary for government or the project authorities to give plausible and acceptable explanations for their project approach and create public awareness about the same. Whereas the government has control over the public funds it has also ventured into the role of a facilitator under liberalisation. Here is an opportunity. Most of the work where public interface comes can be channeled through private agencies and NGOs. They should in turn be asked to form appropriate linkage with the beneficiary stakeholders and the NGOs in this sector. To attain their commercial target the private entrepreneurs will find appropriate local level interventions.

The government's efforts for common and uniform resettlement and rehabilitation has not been successful as different states have different sets of rules and acts. As a first step the government can evolve a common guidelines for hydro project implementation, rehabilitation and resettlement of project site oustees. These guidelines should be uniformly applicable in all states. The guidelines may be made public.

All this will succeed only when the government has a uniform resettlement and rehabilitation policy (same in all states) that is people centred, transparent and perceptibly beneficial. Objections of NGOs should be studied and their solutions or counter arguments should be made public. Public Awareness of 'water security' at the national level; and need for exploitation of locally available, economically and technically viabile sites will be necessary. The R&R plan should really emerge from beneficiaries or be acceptable to them.

The ultimate solution to the opposition to large dams in India can never be found through a confrontationist attitude but can only be arrived by having a continuous and meaningful dialogue with the stakeholder beneficiaries and NGOs. 'Project development' and 'public dialogue' will have to go hand in hand.

The onus for initiating and continuing such a dialogue lies with the government and policy makers who decide on approach for public fund utilisation.

ANNEXURE I

SUSTAINABLE DEVELOPMENT OF POWER SECTOR IN INDIA

Power is a critical infrastructure for the development of the economy and living standards of the citizens. Unlike most other commodities, electricity cannot be stored for future use. Therefore, the generation and consumption has to be done simultaneously. This peculiar feature of electricity makes the dynamics of demand and supply difficult to manage.

Growth of Installed Power Capacity

At the outset, I may point out that the achievement of increasing installed power capacity from a meagre 1326 MW to over 100,000 MW since independence (1947) and electrification of more than 5 lakh villages by all means is nothing but impressive in absolute terms. However, it is a matter of concern that annual per capita consumption of India, at about 350 kWh, is amongst the lowest in the world. The per capita consumption in Brazil is 1,783, China 719, UK 5843, Australia 6606 and USA 8,747 kWh. Further, people in a large number of villages have no access to electricity. The end-users like households, farmers, commercial establishments, industries are confronted with frequent power cuts, both scheduled and unscheduled. Power cuts, erratic voltage and low or high supply frequency have added to the 'power woes' of the consumers.

Power Situation — Lack of Quality, Security and Reliability

The present power situation, lacks not only in terms of performance but also quality, security and reliability. In spite of impressive growth in the early decades of planning, the present total installed capacity of over 102,000 MW is still inadequate to meet our demand. The energy

---

Platinum Jubilee Lecture by Shri Santosh Mohan Deb, M.P., Chairman Parliamentary Committee on Energy organised by the Institution of Engineers (India), Delhi State Centre on November 27, 2002 at New Delhi.

and peak power shortages are reported to be at the level of 12 per cent and 8 per cent respectively. It is estimated that the future additional power requirements will be around 1,40,000 MW at a cost of Rs. 6,00,000 crores to be installed by 2012. This amount of huge investment, the Government feel, would be beyond its reach. This implies that both the Government and the private sector will have to bring in foreign investors with latest technology. Along with the investments, efforts will have to be made to reduce the present high cost of per MW addition of generation capacity. Major part of our generation capacity is coal based. But, high cost of coal, its poor quality and constantly increasing transportation cost add up to high cost of additional capacity generation. The greater ash contents in the coal lead to environmental problems which are also to be dealt with by the thermal power stations adding up to the cost of new capacity.

To supplement the efforts of the States in bridging the gap between demand and supply it was decided in the middle of 1970s to set up generation stations and associated high/extra voltage transmission lines in the central sector. Today, states control about 60 per cent of the country's generation capacity, 70 per cent of the transmission network and 100 per cent of the distribution.

Need to Support Hydro-electricity Generation

For the sustainable development of Power Sector, the contribution of Hydro-Power, need not be over-emphasised. Hydroelectricity is clean energy and its generation is not linked to issues concerning fuel supply. Less than one-fourth of the vast hydel potential of 1,50,000 MW has been tapped so far. When compared to the high utilisation of hydro potential in countries like Norway (58 %), Canada (41 %) and Brazil (31 %), the utilisation of only 17 per cent of its hydel potential by India is extremely low. In fact, the share of hydro-generation in India has gradually declined during the past 25 years. Consequently, thermal generation, which should generally be used for base load operation, is also being used to meet peaking requirements. As against the desirable hydro share of 40 per cent the current is only about 25 per cent in the country. In early 50s the hydel: thermal ratio was 33:67. Since then, there is continuous adverse hydro/thermal mix and now has reached alarming 25:75.

Mismatch in Investment in Generation and Transmission and Distribution

Another weakness encountered in the development of power sector is mismatch in the investment between generation and transmission/ distribution. Inadequate investment in transmission and distribution infrastructure have resulted in power evacuation constraint from the generation stations. The problem has been severe in the eastern region. As against, thumb rule of 1:1 investment in generation and transmissions the investment has been lopsided and ranges in the, reverse ratio of 2:1. It is estimated that for building over 1,00,000 MW of additional power capacity and associated transmission and distribution infrastructure, nearly Rs 8,00,000 crores of investment would be needed in the next decade. The investors have been wary of the sector due to lack of confidence in getting returns on their investment. The payment security measures taken till now have not yielded desired results. There is little doubt that resource generation with in the sector through prompt and efficient collection of appropriate user charges from all the electricity consumers is only a long-term solution to attract investment in the sector. The sector has to be made financially strong from within to attract investments from outside.

Tariff not Related to Cost

Other inherent weaknesses of the power sector include tariffs not related to costs of operation, the inefficient operational phases and nearly 50 per cent of the energy consumed not metered which go towards agriculture consumption, hut lighting, T&D losses and pilferage. T&D losses reported by many SEBs are fudged figures. There is free or subsidised power supply and absence of commercial outlook. Shortage of power and energy is perennial. There was lack of clear-cut policies, organisational purpose, control or responsibility and frequent change of leadership. This is coupled with over staffing and low productivity; and revenue earning distribution function is totally neglected. The poor financial health of SEBs resulted in non-payment of dues to another government companies producing coal. It also affected the capacity of SEBs to properly maintain their plants resulting in very low Plant Load Factor. This in turn led to low generation and huge gap in demand and supply of electricity. This forced the Government to have a re-look on the affairs of the power sector and the Government started the process of restructuring the power sector throughout the country.

SEB Reform

During 1990s the states were asked to unbundle their SEBs into small manageable and efficient corporation, which attract private investment to meet the power sector needs. The private sector was hesitant to invest in the sector because of poor financial condition of most of the SEBs in the country. They were not sure of returns from and safety of their investment. Even the international organisations were hesitant to give loans in the existing scenario. It was suggested/desired by these organisations that they would lend money only when SEBs become financially viable by undertaking a set of reforms. Hence the State Governments have now started the necessary reforms by unbundling SEBs and by putting in place the regulatory mechanism in their states. To help the State Governments in the reform process, the Union Government issued policy guidelines from time to time and carried out amendments in the three Central Acts which presently govern the power sector in the country.

Electricity Bill, 2002

A comprehensive Bill called the Electricity Bill, 2002, covering the entire power sector, has been introduced in the Lok Sabha. The same is now being examined by the Parliamentary Standing Committee on Energy. The Committee has almost finalised its Report and is going to present the Report shortly. The Bill replaces the existing Acts governing the electricity industry in the country.

The Bill envisages bold and pragmatic solutions to salvage the Power Sector from the present situation. Some of the worth mentioning provisions of the Bill are as under:

(i) Thermal generation has been totally de-regulated and is not subjected to the scrutiny of the CEA. However, on the consideration of inter-State dimensions the hydel projects are required to be apprised by CEA.

(ii) The role of CEA has been reoriented. It is now the main technical adviser of Centre and State Governments and Regulators.

(iii) Captive generation is made totally free. Captive generation would also have open access through the grid.

(iv) Generation from non-conventional sources of energy is to be promoted. It is proposed that at least 10 per cent of power be secured from non-conventional energy sources.

(v) There will be government control over Load Despatch Centres, which is critical for the purpose of grid discipline. Load Despatch Centres have been barred from trade in power.

(vi) There will be transmission utility at Centre and State levels. Government companies have been charged with the responsibility for coordination and planned development of transmission network.

(vii) Private sector participation in transmission will be encouraged.

(viii) There will be non-discriminatory open access in transmission. There will be a surcharge to take care of current subsidy. The surcharge will be progressively reduced. Open access is allowed in distribution in phases.

(ix) State Government will have to provide subsidy if it wishes the tariff for a particular class of consumers to be lowered than prescribed.

(x) Metering would be made mandatory.

(xi) Rural Electrification is to be encouraged. The management of local distribution in rural areas permitted through users' associations, cooperatives, panchayati institutions and other local bodies.

(xii) Stringent punishment is possible for theft of power.

(xiii) Regulatory Commission made compulsorily.

(xiv) Trading in power promoted and permitted as a distinct activity with licence.

(xv) Role of Government reduced to making national policies and plans including tariff.

(xvi) State Government given flexibility either to unbundle SEBs or to retain the present monolithic structure.

(xvii) Emphasis on the protection of consumers interest — penalty in the event of failure of giving connection, no sum due from consumer recoverable after a period of two years, consumer to be given connection within one month of giving application, redressal forum for consumers' grievances to be appointed by each distribution licence, etc.

I am optimistic that with the enactment of this Bill into an Act the Power Sector will regain its lost glory. Eventually, it will drive the economy, to newer heights. I look forward for such an opportunity.

ANNEXURE II

ENERGY FOR PEOPLE, ENERGY FOR PEACE: MARKETS, SUSTAINABILITY AND DRIVERS

I am very pleased to join this important gathering to share the work of World Energy Council with you. WEC has member committees in 96 countries, including India which hosted our World Congress back in 1983 and our Annual Executive Assembly meeting in 2000, and I want to make sure the upstream and midstream oil and gas companies are fully plugged into our work on sustainable energy systems.

Up behind me you will see a break out of the energy goals, key issues and linkages which the energy business, government and regulatory leaders debated at the World Energy Congress in Buenos Aires in October 2001. Some of you were probably there because you work for energy companies, you advise governments or you write regulations.

The three areas I want to cover today have a direct impact on your strategic planning and your prospects: the Death of Ideology in Market Reform; the Focus on Sustainable Development; and, New Thinking on Energy Drivers to 2050. All three lead to the conclusion that energy will get more expensive in real terms in the coming years.

Market Reform: The Death of Ideology

The ideology of market reform, often equated with deregulation, liberalisation or privatisation, seems to have gone out the window with California, ENRON, and Brazil. I won't spend time on an analysis of California or what happened in ENRON because you have been to many conferences already held on these topics. Now, in the UK, perhaps one of the most liberalised energy markets in the world, there is the danger that British Energy (which is the largest electricity supplier, mainly from nuclear power) will go bankrupt because of barriers to exits even though wholesale electricity prices have fallen 40%! *Lets just say that markets fail and regulations can be non-existent or, what is worse, misguided.*

Platinum Jubilee Lecture by Shri Gerald Doucet, Secretary General, World Energy Council (UK), organised by the Institution of Engineers (India), Delhi State Centre on November 27, 2002 at New Delhi.

Brazil's energy crisis last year doesn't get much analytical ink: faced with chronic shortages of water to feed its 95 per cent reliance on hydro electricity, otherwise called putting too many energy eggs in one basket, the leadership in Brazil established price signals and rewards for a significant cut in energy consumption within a six-month period. In Brazil, the fundamental market question is how to get sufficiently strong price signals to motivate investment in new capacity and diversified energy sources to meet growing electricity demand.

Yet, one must be careful not to throw the market reform baby out with the regulatory bath water. What California and ENRON show is that *the ideology of free markets in energy services is not sufficient because careful market design and effective regulatory regimes are also essential.* It is worth noting that not one end-user failed to receive a kWh of electricity or a joule of gas as a result of the ENRON failure; this is the single biggest endorsement of the market in recent times. But it also shows that energy is not a commodity like any other for, when the lights go out, or diesel or natural gas prices rise rapidly, politicians often take serious action.

Market reform, to promote competitive pricing, technological innovation, customer choice and trade in energy services, coupled with appropriate regulation to address *fairness and sustainability issues* (such as capacity, reliability, affordability and environmental goals), is the first priority of energy policy. *Keeping energy options open in face of growing demand for electricity is a close second.*

These priorities have been under attack in Europe and Asia, not to mention in India where the opening of the electricity market has been put on hold. The focus in some countries on eliminating nuclear power from the energy mix or establishing costly renewables quotas in electricity generation is regrettable. *The wide political endorsement of the Kyoto Protocol without adequate debate of its fullest economic trade-offs has raised false hopes about the global governance of GHG emissions.*

So, when it comes to market reform, we have to do a much better job about the design and application of sound regulations, the range of which is probably going to be wider than we might have expected just a year or so ago. In terms of customer choice, trade in energy services, and third party access, market reform is not the culprit of sustainable energy systems; it is the heart and soul. But not enough time is put on the regulations needed to achieve the objectives of universal access, capacity and quality of service, and acceptability in terms of social and environmental externalities. Interestingly enough, our *Congress in October 2001 focused on investor-friendly regulation* and *the bottlenecks caused by inadequate transmission infrastructure or interconnectivity for both gas and electricity.* WEC is working on these issues in a major new piece of research to be released at our Sydney Congress in 2004.

With the death of ideology in market reform, business and government people are coming together to focus on *competition-friendly regulation*, which addresses the three pillars of *sustainable energy development: the economics, the social, and the environmental.* While market reform is putting downward pressure on wholesale and retail energy prices in some big industrialised markets, it results in higher prices in others. A broader range of appropriate regulations, leading to such things as cleaner technologies, full fuel costs accounting and emissions trades is likely to lead to higher prices overall.

WEC is a non-governmental, global multi-energy body and it sees this challenge in business terms, in terms of security and sustainability, not in moralistic terms, although there is plenty of evidence that *good business and sound ethics* are two sides of the same coin. We are not alone in this thinking. Matthew Simmons, an oil and gas analyst recently published a speech on Energy and Water at the 15th Camden Conference, which I urge you to read on his website. He says: "A good question to ponder is why something so precious (as energy) is so misunderstood or even reviled by some very well-intentioned and well-educated people. There are numerous surveys that confirm that many people are suspicious about most energy sources. Unless energy prices are cheap, people tend to think that they are "rigged". Energy companies are often accused of encouraging wasteful energy use in order to pad their balance sheets. Our environmental movement which essentially began in the wake of the 1969 Santa Barbara Oil Spill 33 years ago, instantly targeted oil and gas as their public enemy "Number One". It has never relented in this battle, despite remarkable environmental improvements made by most energy sources over the last 30 years."

The Focus on Sustainability

I have just come from the World Summit on Sustainable Development in South Africa, a gathering of 60,000 international officials, business people, academics, NGOs, and 140 governments.

I think the oil and gas industry can take comfort from what we achieved when it comes to energy. We have made progress in showing that market reform and keeping all energy options open are part of the solution to poverty eradication.

Unlike Rio ten years ago, nobody in Johannesburg was buying the early demise of fossil fuels or nuclear power in the world's energy mix; there were NGOs who made their disappointment known but they were put in their place by the developing countries themselves, where two billion people do not have access to any commercial energy at all, with great harm to their health, human rights, and the environment. Nobody in Johannesburg was focused exclusively on the environment; access to commercial energy as part of a global effort to eradicate poverty was the number one issue. And that is why foolish efforts to set quotas in energy supply for renewables were soundly defeated.

But lets put sustainability into a business context. *Two big energy policy statements* by the USA and by the European Union have been made in the last year. *Both are supply focused*, although the EU does address demand management. Both imply Crises when, in fact, these important regions face Choices. If you want to see an energy crisis, when it comes to supply, India has 51 per cent technical and non-technical losses in its 90,000MW grid and needs another 25,000 MW of affordable capacity by 2005. If you want to see a crisis when it comes to energy demand, go to Africa where, in the space of ten years of debate on opening up markets to attract new investment in energy infrastructure, the proportion of the total population with a minimum amount of electricity has fallen from 12 per cent to 8 per cent.

Supply and security focused discussion of energy priorities (mainly for oil but also now natural gas and electricity) is, therefore, only part of the story and, for the WEC, not the main plot. In terms of best practices in existing power plants, cleaner technologies, and end-use (especially transportation), fossil fuels will continue to be the major source of energy in India. Energy accessibility in economies in transition and in developing countries (especially Africa) has huge implications for security of supply.

Everybody knows markets don't always work; people pay a price for imperfections and somebody gets the rent. Yet, I hardly think that the one-third of the world, which does not now have access to commercial energy of any kind, is a mere imperfection. It is a disaster! It is at the core of poverty in the world, and we all know poverty can breed terrorism or foolish government policy to satisfy the strongest voices of despair.

It is clear that an excessive focus on lower energy prices—as the desired and expected result of market reform, without any attention to reliability or quality of service issues—*is misleading and does a disservice to the debate on market reform*; indeed, lower energy prices come about only in markets where there is overcapacity and enough players to achieve reasonable competition; that certainly doesn't apply in many developing countries. Somebody has to pay for improved access and for reliability, and I think it will be the consumer, wherever he resides. *Sustainable energy systems do not come free! The role of subsidies in making energy affordable in the poorest countries in the world is a key issue.*

WEC doesn't have the answers to all of these questions. WEC work going back to our Millennium Statement on Energy for Tomorrow's World-Acting Now! argued that *keeping all energy options open, trade in energy services on a regional if not a global level*, and *appropriate regulations to govern these as well as emissions targets were key pillars of sustainable energy development.* We set a target of 500 kWh per person per year for everyone in the world by 2020; that's a light bulb and a fridge! We estimated that $30 billion per annum in financing 200GW of new capacity over 20 years, in addition to current maintenance and upgrade investments of about $240 billion per year would do the trick! We produced a study on Pricing Energy in Developing Countries last year which attempted to revise recent thinking on subsidy capture and marginal cost pricing to drive the necessary investment. *We have done major casework on the Ethical Dimension of the energy business, on electricity market design, on energy markets in transition, and on energy R&D expenditures and new generation technologies.* You can visit all this work on our global energy information system at www.worldenergy.org.

Energy Drivers for the 2050 Market

At our Executive Assembly in Buenos Aires the WEC approved a new study on Energy Drivers looking out to 2050 under the chairmanship of Dr. Majid Al-Moneef, who is economic adviser to the Minister of Petroleum of Saudi Arabia. Together, with the help of a Study Group of over 40 experts from WEC Member Committees around the world, they have questioned conventional wisdom and begun the process of pulling together a report which will guide four other WEC studies on market reform, technologies, lifecycle analysis, and climate change.

I thought I would give you a snapshot of the work to date.

Macroeconomic Trends

WEC is concerned that in the hundreds of supply-demand scenarios out there, including a few done by WEC and IIASA to 2100 (after all, the inertia of energy systems means a big lead time for change, or something like that), the assumptions behind them are outmoded or feedbacks are not adequately taken into account. We held a special symposium on macroeconomic trends at Chatham House in London in February to smoke out the latest thinking.

Our study confirms the growth of the three factors of population, GDP and Total Primary Energy Requirements, but a growth that progressively loses its momentum as we approach 2050. The demographic transition is now over, and the world population is progressively aging, possibly peaking at 8 billion people in the 2040s before declining. *In developed countries, energy demand is now decoupled from economic growth and the same process may soon begin to pass through leading developing country markets.*

WEC is moving towards the view that, *in terms of global energy, the likely evolution will be characterised increasingly by stop and go episodes, each episode exhibiting a decline in energy prices followed by a significant rise.* One does not know when these shocks will occur, or what their impacts or feedbacks will be in terms of slowing GDP, lowering energy intensities, or changing the energy mix, but the days of extrapolated linear trends may have come to an end.

Supply Trends

The WEC held a special symposium on supply side issues in Ankara in April, involving Middle East, Central European, Asian and American experts.

With the rebound for coal after the first oil shock of 1973, *coal stopped its secular decline in terms of primary energy supply and maintained its market share at about 25 per cent.* WEC suspects the decline of coal, despite recent positive developments in terms of efficiency and cleaner technologies, will resume, not as fast as the recent period led by the Chinese shrinkage, but it will happen slowly and regularly, unless there is a technological breakthrough in terms of commercial zero emission coal plants which will put the cost of coal-based electricity on a par with that of nuclear power. Yet, despite the gradual erosion of its market share, coal production will continue to rise until the 2030-40 period, after which there will be that an absolute decline in its share of total primary energy supply will take hold.

On natural gas it is WEC's view that the impressive inroads of this energy source in *the world's energy supply will not be maintained all the way to 2050*. Unless US federal lands are opened up to exploration and unless global natural gas prices rise significantly, there will be no magic potion to alter the coming plateau and eventual decline of natural gas supplies in North America and some other key markets. Whether natural gas prices can be decoupled from oil prices, with the rise in prices in North America and growing interest in LNG trade, and whether the volatility of natural gas and oil markets can be effectively managed over the longer term, are factors which the WEC is now studying. Unconventional gas could remain a stable source for a number of years, but the overall size of this source at higher natural gas price levels does not offset the overall decline in US gas production. WEC is currently assessing data on Mexico and, of course, Europe, Russia, Middle East, Central Asia, and Africa.

It will take higher natural gas prices to bring reserves into new production but, at the same time, these higher prices will encourage a switch to other energy sources. The need for new infrastructure to deliver natural gas to key markets will also add to the upward price pressure. The role of the emerging GASPEC might also be of considerable importance in all of these aspects, even though I suspect that gas, with no long-term captive market, will be less prone to a cartel or market power than oil has been.

Central to the WEC work on Drivers is not so much what the reserve base for oil, coal or natural gas is but, rather, where these reserves are and how much it will take in terms of higher prices for energy (from diesel fuel to electricity) to trigger the investments in production and infrastructure to overcome the regional and global bottlenecks between supply and demand. In other words, WEC's concern is not the size of the tank but the size of the tap; this is most true for oil. *WEC believes there are two new drivers, when it comes to oil: first, if one excludes the important but short-lived results in deep-water production, non-OPEC oil production reached a plateau in 1997; second, OPEC might face an unsustainable call on its oil both from within its markets and abroad.* Both of these Drivers point to a possible new break in the long-term downward trend in real oil prices.

From its early period, the oil market was very volatile unless it was managed by a dominant actor or actors. This feature results from a combination of high fixed costs calling for high prices to launch new investments and for full utilisation after new discoveries come on stream, and low variable costs pushing prices down as long as over capacity exits. This interplay of cost pressures is one of the primary forces in the stop and go episodic impacts on energy prices which the WEC foresees.

Of course, if fossil fuel prices rise in the shorter term either because of the cost of new, cleaner technologies or because of regulations to internalise CO_2 waste management costs in the price of energy services based on fossil fuels, both renewables and nuclear power will benefit, in relative terms. Nuclear power is stuck on a plateau, but growing awareness of rational criteria regarding the economics, security/ reliability and environmental performance favour a significant rebirth of nuclear power, even in some developing countries, when present power plants have to be replaced in the 2010-20 period.

The lead renewable is hydropower which is expected to maintain its current share of total energy supply before tapering off in 2030-40. The share of non-commercial fuels (mostly biomass) will only grow slowly before declining around 2030. New modern renewables (including commercial biomass) will show surprising growth rates from a low base, particularly in hybrid systems linked to baseload electricity, but they will continue to contribute only a small share of world energy supply by 2030 and probably later because of their dispersed and intermittent nature.

Adjusting energy needs to supply is both a threat and an opportunity. It is a threat for the poorest countries and for those who do not now have access to modern energy because the upward adjustment in prices that the WEC foresees might further delay affordable access. It is an opportunity because a number of energy policies, based on cheap energy but unsustainable energy, will have to be revised, and vested interests of some powerful lobbies will give way to more appropriate energy regulations that underpin competition, metering, collection, and effective subsidy capture.

Demand Trends

WEC has just completed its special symposium on demand trends and energy efficiency issues in Mexico city. We are still absorbing all that was said. We have found that consumers behave rationally, some over-consuming energy and some under-consuming. But aggregate consumer behaviour closely follows the level of final prices. Energy efficiency policies do not seem to impact energy consumption as long as they do not alter final energy prices, either directly through taxes or indirectly through binding regulation. In countries with hidden price distortions (such as non-payment for company cars or district heating), the constancy of energy trends over time suggests that such distortions have not been offset by energy efficiency policies.

The WEC is doing some new work on energy-related services as part of its work on Demand Trends. So far we see that, at a given income/GDP level, price is the most important factor by far in terms of consumption patterns; final price elasticity for energy services seems to be close to one, in the sense that the ultimate objective of the consumer is a constant energy budget. Most importantly, but worth further analysis, when real energy prices change, the break in energy demand trends lasts two years before energy efficiency compensations move the consumption back onto the trend line. That is not a long time in the slow life of energy systems! *The implication of this is that the new technologies and appliances, brought to market because of higher energy prices, result in higher economic welfare rather than lower energy demand.*

Increased energy efficiency seems to mean more energy demand not less, with a time lag, of course. The snowballing of energy demand which can be seen to repeat itself many times over the last two centuries was only stopped by price hikes caused by external events such as wars, drastic changes in market rules, or temporary incapacity of the rich resource base to supply exponentially increasing energy demand. Extrapolation makes little sense in this kind of stop and go process, not only because the breaks come as surprises but also because they trigger through price increases a number of technological and behavioural reactions that cannot be predicted.

Finally, the WEC is moving towards the view that, whether we like it or not, oil has been and will continue to be the price setter with enormous implications for all energies. And when you talk oil prices you need quickly to focus on oil availability in the Middle East, particularly Saudi Arabia. This is not to say natural gas, coal, renewables and nuclear power are not important; but, make no mistake, a shortage in any of these other energies will result in an increased call on oil directly or indirectly.

Conclusions

All of this work leads to two central questions of the day on which energy policy and business people around the world should dwell: first, will OPEC have sufficient idle capacity to be able to supply a corresponding surge of oil demand with a smooth evolution to higher real oil prices? *And, second,* is the true meaning of the War on Terrorism and the new presence of America in Central Asia republics and former Soviet states that Russia and USA now have a new energy partnership to delay or counterbalance Middle East influence with faster and larger shipments of Russian oil and natural gas to world markets?

My answer to the first question is YES; the work of the Study Group is revealing clear insights about the positive impacts of OPEC since it was created in 1960. My answer to the second question is also YES, not because Russian or Caspian oil and gas will drastically change the economics of the world energy system but because oil is truly a matter of international politics.

In a world where it might become harder, not easier, to bring affordable commercial energy access to every person in this world, better understanding among producers and consumers of energy of oil and natural gas is a precious asset. Harder it may be, but we must achieve it sooner rather than later if we want to live in a safe and sustainable world.

Thank you.

ANNEXURE III

WORLD SUMMIT ON SUSTAINABLE DEVELOPMENT

PLAN OF IMPLEMENTAION

The Johannesburg Declaration on Sustainable Development

From our Origins to the Future

1. We, the representatives of the peoples of the world, assembled at the World Summit on Sustainable Development in Johannesburg, South Africa from 2-4 September 2002, reaffirm our commitment to Sustainable development.
2. We commit ourselves to build a humane, equitable and caring global society cognisant of the need for human dignity for all.
3. At the beginning of this Summit, the children of the world spoke to us in a simple yet clear voice that the future belongs to them, and accordingly challenged all of us to ensure that through our actions they will inherit a world free of the indignity and indecency occasioned by poverty, environmental degradation and patterns of unsustainable development.
4. As part of our response to these children, who represent our collective future, all of us, coming from every corner of the world, informed by different life experiences, are united and moved by a deeply-felt sense that we urgently need to create a new and brighter world of hope.
5. Accordingly, we assume a collective responsibility to advance and strengthen the interdependent and mutually reinforcing pillars of Sustainable development — economic development, social development and environmental protection — at local, national, regional and global levels.
6. From this Continent, the Cradle of Humanity we declare, through the Plan of Implementation and this Declaration, our responsibility to one another, to the greater community of life and to our children.
7. Recognising that humankind is at a crossroad, we have united in a common resolve to make a determined effort to respond

positively to the need to produce a practical and visible plan that should bring about poverty eradication and human development.

From Stockholm to Rio de Janeiro to Johannesburg

8. Thirty years ago, in Stockholm, we agreed on the urgent need to respond to the problem of environmental deterioration. Ten years ago, at the United Nations Conference on Environment and Development, held in Rio de Janeiro, we agreed that the protection of the environment, and social and economic development are fundamental to sustainable development, based on the Rio Principles. To achieve such development, we adopted the global programme. Agenda-21, and the Rio Declaration, to which we reaffirm our commitment. The Rio Summit was a significant milestone that set a new agenda for sustainable development.
9. Between Rio and Johannesburg the world's nations met in several major conferences under the guidance of the United Nations, including the Monterrey Conference on Finance for Development, as well as the Doha Ministerial Conference. These conferences defined for the world a comprehensive vision for the future of humanity.
10. At the Johannesburg Summit we achieved much in bringing together a rich tapestry of peoples and views in a constructive search for a common path, towards a world that respects and implements the vision of sustainable development. Johannesburg also confirmed that significant progress has been made towards achieving a global consensus and partnership amongst all the people of our planet.

The Challenges we Face

11. We recognise that poverty eradication, changing consumption and production patterns, and protecting and managing the natural resource base for economic and social development are overarching objectives of, and essential requirements for sustainable development.
12. The deep fault line that divides human society between the rich and the poor and the ever-increasing gap between die

developed and developing worlds pose a major threat to global prosperity, security and stability.

13. The global environment continues to suffer. Loss of biodiversity continues, fish stocks continue to be depleted, desertification claims more and more fertile land, the adverse effects of climate change are already evident, natural disasters are more frequent and more devastating and developing countries more vulnerable, and air, water and marine pollution continue to rob millions of a decent life.
14. Globalisation has added a new dimension to these challenges. The rapid integration of markets, mobility of capital and significant increases in investment flows around the world have opened new challenges and opportunities for the pursuit of sustainable development. But the benefits and costs of globalisation are unevenly distributed, with developing countries facing special difficulties in meeting this challenge.
15. We risk the entrenchment of these global disparities and unless we act in a manner that fundamentally changes their lives, the poor of the world may lose confidence in their representatives and the democratic systems to which we remain committed, seeing their representatives as nothing more than sounding brass or tinkling cymbals.

Our Commitment to Sustainable Development

16. We are determined to ensure that our rich diversity, which is our collective strength, will be used for constructive partnership for change and for the achievement of the common goal of sustainable development.
17. We welcome the Johannesburg Summit focus on the indivisibility of human dignity and are resolved through decisions on targets, timetables and partnerships to speedily increase access to basic requirements such as clean water, sanitation, energy, health care, food security and the protection of bio-diversity. At the same time, we will work together to assist one another to have access to financial resources, benefit from the opening of markets, ensure capacity building, use modern technology to bring about development, and make sure that there is technology transfer, human resource development, education and training to banish forever underdevelopment.

18. We are committed to ensure that women's empowerment and emancipation, and gender equality are integrated in all activities encompassed within Agenda-21, the Millennium Development Goals and the Johannesburg Plan of Implementation.
19. We recognise the reality that global society has the means and is endowed with the resources to address the challenges of poverty eradication and sustainable development confronting all humanity. Together we will take extra steps to ensure that these available resources are used to the benefit of humanity.
20. In this regard, to contribute to the achievement of our development goals and targets, we urge developed countries that have not done so to make concrete efforts towards the internationally agreed levels of Official Development Assistance.
21. We welcome and support the emergence of stronger regional groupings and alliances, such as the New Partnership for Africa's Development (NEPAD), to promote regional co-operation, improved international co-operation and promote sustainable development.
22. We shall continue to pay special attention to the developmental needs of Small Island Developing States and the Least Developed Countrics.
23. We recognise sustainable development requires a long-term perspective and broad-based participation in policy formulation, decision-making and implementation at all levels. As social partners we will continue to work for stable partnerships with all major groups respecting the independent, important roles of each of these.
24. We agree that in pursuit of their legitimate activities the private sector, both large and small companies, have a duty to contribute to the evolution of equitable and sustainable communities and societies.
25. We also agree to provide assistance to increase income generating employment opportunities, taking into account the International Labour Organisation (ILO) Declaration of Fundamental Principles and Rights at Work.
26. We agree that there is a need for private sector corporations to enforce corporate accountability. This should take place

within a transparent and stable regulatory environment.

27. We undertake to strengthen and improve governance at all levels, for the effective implementation of Agenda-21, the Millennium Development Goals and the Johannesburg Plan of Implementation.

Multilateralism is the Future

28. To achieve our goals of sustainable development, we need more effective, democratic and accountable international and multilateral institutions.
27. We reaffirm our commitment to the principles and purposes of the UN Charter and international law as well as the strengthening of multi-lateralism. We support the leadership role of the United Nations as the most universal and representative organisation in the world, which is best placed to promote sustainable development.
28. We further commit ourselves to monitor progress at regular intervals towards the achievement of our sustainable development goals and objectives.

Making it Happen!

29. We are in agreement that this must be an inclusive process, involving all tlie major groups and governments that participated in the historic Johannesburg Summit.
30. We commit ourselves to act together, united by a common determination to save our planet, promote human development and achieve universal prosperity and peace.
31. We commit ourselves to the Johannesburg Plan of Implementation and to expedite the achievement of the time-bound, socio-economic and environmental targets contained therein.
32. From the African continent, the Cradle of Humankind, we solemnly pledge to the peoples of the world, and the generations that will surely inherit this earth, that we are determined to ensure that our collective hope for sustainable development is realised.

We express our deepest gratitude to the people and the Government of South Africa for their generous hospitality and excellent arrangements made for the World Summit on Sustainable Development.

INTRODUCTION

1. The United Nations Conference on Environment and Development (UNCED), held in Rio de Janeiro in 1992, provided the fundamental principles and the programme of action for achieving sustainable development. We strongly reaffirm our commitment to the Rio principles, the full implementation of Agenda-21 and the Programme for the Further Implementation of Agenda-21. We also commit ourselves to achieving the internationally agreed development goals, including those contained in the United Nations Millennium Declaration and in the outcomes of the major United Nations conferences and international agreements since 1992.

2. The present plan of implementation will further build on the achievements made since UNCED and expedite the realization of the remaining goals. To this end, we commit ourselves to undertaking concrete actions and measures at all levels and to enhancing international cooperation, taking into account the Rio Principles, including, inter alia, the principle of common but differentiated responsibilities as set out in principle 7 of the Rio Declaration on Environment and Development. These efforts will also promote the integration of the three components of sustainable development — economic development, social development and environmental protection — as interdependent and mutually reinforcing pillars. Poverty eradication, changing unsustainable patterns of production and consumption, and protecting and managing the natural resource base of economic and social development are overarching objectives of, and essential requirements for, sustainable development.

3. We recognize that the implementation of the outcomes of the Summit should benefit all, particularly women, youth, children and vulnerable groups. Furthermore, the implementation should involve all relevant actors through partnerships, especially between Governments of the North and South, on the one hand, and between Governments and major groups, on the other, to achieve the widely shared goals of sustainable development. As reflected in the Monterrey Consensus, such partnerships are key to pursuing sustainable development in a globalizing world.

4. Good governance within each country and at the international

level is essential for sustainable development. At the domestic level, sound environmental, social and economic policies, democratic institutions responsive to the needs of the people, the rule of law, anticorruption measures, gender equality and an enabling environment for investment are the basis for sustainable development. As a result of globalization, external factors have become critical in determining the success or failure of developing countries in their national efforts. The gap between developed and developing countries points to the continued need for a dynamic and enabling international economic environment supportive of international cooperation, particularly in the areas of finance, technology transfer, debt and trade, and full and effective participation of developing countries in global decision-making, if the momentum for global progress towards sustainable development is to be maintained and increased.

5. Peace, security, stability and respect for human rights and fundamental freedoms, including the right to development, as well as respect for cultural diversity, are essential for achieving sustainable development and ensuring that sustainable development benefits all.

5(a). We acknowledge the importance of ethics for sustainable development, and therefore we emphasize the need to consider ethics in the implementation of Agenda-21.

POVERTY ERADICATION

6. Eradicating poverty is the greatest global challenge facing the world today and an indispensable requirement for sustainable development, particularly for developing countries. Although each country has the primary responsibility for its own sustainable development and poverty eradication and the role of national policies and development strategies cannot be overemphasized, concerted and concrete measures are required at all levels to enable developing countries to achieve their sustainable development goals as related to the internationally agreed poverty-related targets and goals, including those contained in Agenda-21, the relevant outcomes of other United Nations conferences and the United Nations Millennium Declaration. This would include actions at all levels to:

(a) Halve, by the year 2015, the proportion of the world's people whose income is less than $1 a day and the proportion of people who suffer from hunger and, by the same date, to halve the proportion of people without access to safe drinking water;

(b) Establish a world solidarity fund to eradicate poverty and to promote social and human development in the developing countries pursuant to modalities to be determined by the General Assembly, while stressing the voluntary nature of the contributions, the need to avoid duplication of existing United Nations funds, and encouraging the role of the private sector and individual citizens relative to Governments in funding the endeavours;

(c) Develop national programmes for sustainable development and local and community development, where appropriate within country-owned poverty reduction strategies, to promote the empowerment of people living in poverty and their organizations. These programmes should reflect their priorities and enable them to increase access to productive resources, public services and institutions, in particular land, water, employment opportunities, credit, education and health;

(d) Promote women's equal access to and full participation, on the basis of equality with men, in decision-making at all levels, mainstreaming gender perspectives in all policies and strategies, eliminating all forms of violence and discrimination against women, and improving the status, health and economic welfare of women and girls through full and equal access to economic opportunity, land, credit, education and health-care services;

(e) Develop policies and ways and means to improve access by indigenous people and their communities to economic activities, and increase their employment through, where appropriate, such measures as training, technical assistance and credit facilities. Recognize that traditional and direct dependence on renewable resources and ecosystems, including sustainable harvesting, continues to be essential to the cultural, economic and physical well-being of indigenous people and their communities;

(f) Deliver basic health services for all and reduce environmental health threats, taking into account the special needs of children and the linkages between poverty, health and environment, with provision of financial resources, technical assistance and knowledge transfer to developing countries and countries with economies in transition;

(g) Ensure that children everywhere, boys and girls alike, will be able to complete a full course of primary schooling and will have equal access to all levels of education;

(h) Provide access to agricultural resources for people living in poverty, especially women and indigenous communities, and promote, as appropriate, land tenure arrangements that recognize and protect indigenous and common property resource management systems;

(i) Build basic rural infrastructure, diversify the economy and improve transportation and access to markets, market information and credit for the rural poor to support sustainable agriculture and rural development;

(j) Transfer basic sustainable agricultural techniques and knowledge, including natural resource management, to small and medium-scale farmers, fishers and the rural poor, especially in developing countries, including through multi-stakeholder approaches and public-private partnerships aimed at increasing agriculture production and food security;

(k) Increase food availability and affordability, including through harvest and food technology and management, as well as equitable and efficient distribution systems, by promoting, for example, community-based partnerships linking urban and rural people and enterprises;

(l) Combat desertification and mitigate the effects of drought and floods through such measures as improved use of climate and weather information and forecasts, early warning systems, land and natural resource management, agricultural practices and ecosystem conservation in order to reverse current trends and minimize degradation of land and water resources, including through the provision of adequate and predictable financial resources to implement the United Nations Convention to Combat Desertification in Those Countries Experiencing Serious Drought and/or Desertification, particularly in Africa, as one of the tools for poverty eradication; and

(m) Increase access to sanitation to improve human health and reduce infant and child mortality, prioritizing water and sanitation in national sustainable development strategies and poverty reduction strategies where they exist.

7. The provision of clean drinking water and adequate sanitation is necessary to protect human health and the environment. In this respect, we agree to halve, by the year 2015, the proportion of people who are unable to reach or to afford safe drinking water (as outlined in the Millennium Declaration) and the proportion of people who do not have access to basic sanitation, which would include actions at all levels to:

(a) Develop and implement efficient household sanitation systems;
(b) Improve sanitation in public institutions, especially schools;
(c) Promote safe hygiene practices;
(d) Promote education and outreach focused on children, as agents of behavioural change;
(e) Promote affordable and socially and culturally acceptable technologies and practices;
(f) Develop innovative financing and partnership mechanisms; and
(g) Integrate sanitation into water resources management strategies.

8. Take joint actions and improve efforts to work together at all levels to improve access to reliable and affordable energy services for sustainable development sufficient to facilitate the achievement of the millennium development goals, including the goal of halving the proportion of people in poverty by 2015, and as a means to generate other important services that mitigate poverty, bearing in mind that access to energy facilitates the eradication of poverty. This would include actions at all levels to:

(a) Improve access to reliable, affordable, economically viable, socially acceptable and environmentally sound energy services and resources, taking into account national specificities and circumstances, through various means, such as enhanced rural electrification and decentralized energy systems, increased use of renewables, cleaner liquid and gaseous fuels and enhanced energy efficiency, by intensifying regional and international cooperation in support of national efforts, including through capacity-building, financial and technological assistance and innovative financing mechanisms, including at the micro and meso levels, recognizing the specific factors for providing access to the poor;

(b) Improve access to modern biomass technologies and fuelwood sources and supplies, and commercialize biomass operations, including the use of agricultural residues, in rural areas and where such practices are sustainable;

(c) Promote a sustainable use of biomass and, as appropriate, other renewable energies through improvement of current patterns of use, such as management of resources, more efficient use of fuelwood and new or improved products and technologies;

(d) Support the transition to the cleaner use of liquid and gaseous fossil fuels, where considered more environmentally sound, socially acceptable and cost-effective;

(e) Develop national energy policies and regulatory frameworks that will help to create the necessary economic, social and institutional conditions in the energy sector to improve access to reliable, affordable, economically viable, socially acceptable and environmentally sound energy services for sustainable development and poverty eradication in rural, peri-urban and urban areas;

(f) Enhance international and regional cooperation to improve access to reliable, affordable, economically viable, socially acceptable and environmentally sound energy services, as an integral part of poverty reduction programmes, by facilitating the creation of enabling environments and addressing capacity-building needs, with special attention to rural and isolated areas, as appropriate;

(g) Assist and facilitate on an accelerated basis, with the financial and technical assistance of developed countries, including through public-private partnerships, the access of the poor to reliable, affordable, economically viable, socially acceptable and environmentally sound energy services, taking into account the instrumental role of developing national policies on energy for sustainable development, bearing in mind that in developing countries sharp increases in energy services are required to improve the standards of living of their populations and that energy services have positive impacts on poverty eradication and improve standards of living.

9. Strengthen the contribution of industrial development to poverty

eradication and sustainable natural resource management. This would include actions at all levels to:

(a) Provide assistance and mobilize resources to enhance industrial productivity and competitiveness as well as industrial development in developing countries, including the transfer of environmentally sound technologies on preferential terms, as mutually agreed;
(b) Provide assistance to increase income-generating employment opportunities, taking into account the International Labour Organization (ILO) Declaration on Fundamental Principles and Rights at Work;
(c) Promote the development of micro, small and medium-sized enterprises, including by means of training, education and skill enhancement, with a special focus on agro-industry as a provider of livelihoods for rural communities;
(d) Provide financial and technological support, as appropriate, to rural communities of developing countries to enable them to benefit from safe and sustainable livelihood opportunities in small-scale mining ventures;
(e) Provide support to developing countries for the development of safe low-cost technologies that provide or conserve fuel for cooking and water heating;
(f) Provide support for natural resource management for creating sustainable livelihoods for the poor.

10. By 2020, achieve a significant improvement in the lives of at least 100 million slum dwellers, as proposed in the "Cities without slums" initiative. This would include actions at all levels to:

(a) Improve access to land and property, to adequate shelter and to basic services for the urban and rural poor, with special attention to female heads of household;
(b) Use low-cost and sustainable materials and appropriate technologies for the construction of adequate and secure housing for the poor, with financial and technological assistance to developing countries, taking into account their culture, climate, specific social conditions and vulnerability to natural disasters;
(c) Increase decent employment, credit and income for the urban

poor, through appropriate national policies, promoting equal opportunities for women and men;

(d) Remove unnecessary regulatory and other obstacles for microenterprises and the informal sector;

(e) Support local authorities in elaborating slum upgrading programmes within the framework of urban development plans and facilitate access, particularly for the poor, to information on housing legislation.

11. Take immediate and effective measures to eliminate the worst forms of child labour as defined in ILO Convention No. 182, and elaborate and implement strategies for the elimination of child labour that is contrary to accepted international standards.

12. Promote international cooperation to assist developing countries, upon request, in addressing child labour and its root causes, inter alia, through social and economic policies aimed at poverty conditions, while stressing that labour standards should not be used for protectionist trade purposes.

CHANGING UNSUSTAINABLE PATTERNS OF CONSUMPTION AND PRODUCTION

13. Fundamental changes in the way societies produce and consume are indispensable for achieving global sustainable development. All countries should promote sustainable consumption and production patterns, with the developed countries taking the lead and with all countries benefiting from the process, taking into account the Rio principles, including, inter alia, the principle of common but differentiated responsibilities as set out in principle 7 of the Rio Declaration on Environment and Development. Governments, relevant international organizations, the private sector and all major groups should play an active role in changing unsustainable consumption and production patterns. This would include the actions at all levels set out below.

14. Encourage and promote the development of a 10-year framework of programmes in support of regional and national initiatives to accelerate the shift towards sustainable consumption and production to promote social and economic development within the carrying capacity of ecosystems by addressing and, where appropriate, delinking economic growth and environmental degradation through improving

efficiency and sustainability in the use of resources and production processes, and reducing resource degradation, pollution and waste. All countries should take action, with developed countries taking the lead, taking into account the development needs and capabilities of developing countries through mobilization, from all sources, of financial and technical assistance and capacity-building for developing countries. This would require actions at all levels to:

(a) Identify specific activities, tools, policies, measures and monitoring and assessment mechanisms, including, where appropriate, life-cycle analysis and national indicators for measuring progress, bearing in mind that standards applied by some countries may be inappropriate and of unwarranted economic and social cost to other countries, in particular developing countries;
(b) Adopt and implement policies and measures aimed at promoting sustainable patterns of production and consumption, applying, inter alia, the polluter-pays principle described in principle 16 of the Rio Declaration on Environment and Development;
(c) Develop production and consumption policies to improve the products and services provided, while reducing environmental and health impacts, using, where appropriate, sciencebased approaches, such as life-cycle analysis;
(d) Develop awareness-raising programmes on the importance of sustainable production and consumption patterns, particularly among youth and the relevant segments in all countries, especially in developed countries, through, inter alia, education, public and consumer information, advertising and other media, taking into account local, national and regional cultural values;
(e) Develop and adopt, where appropriate, on a voluntary basis, effective, transparent, verifiable, non-misleading and non-discriminatory consumer information tools to provide information relating to sustainable consumption and production, including human health and safety aspects. These tools should not be used as disguised trade barriers;
(f) Increase eco-efficiency, with financial support from all sources, where mutually agreed, for capacity-building, technology transfer and exchange of technology with

developing countries and countries with economies in transition, in cooperation with relevant international organizations.

15. Increase investment in cleaner production and eco-efficiency in all countries through, inter alia, incentives and support schemes and policies directed at establishing appropriate regulatory, financial and legal frameworks. This would include actions at all levels to:

(a) Establish and support cleaner production programmes and centres and more efficient production methods by providing, inter alia, incentives and capacity-building to assist enterprises, especially small and medium-sized enterprises and particularly in developing countries, in improving productivity and sustainable development;
(b) Provide incentives for investment in cleaner production and eco-efficiency in all countries, such as state-financed loans, venture capital, technical assistance and training programmes for small and medium-sized companies while avoiding trade-distorting measures inconsistent with WTO rules;
(c) Collect and disseminate information on cost-effective examples in cleaner production, eco-efficiency and environmental management, and promote the exchange of best practices and know-how on environmentally sound technologies between public and private institutions;
(d) Provide training programmes to small and medium-sized enterprises on the use of information and communication technologies.

16. Integrate the issue of production and consumption patterns into sustainable development policies, programmes and strategies, including, where applicable, into poverty reduction strategies.

17. Enhance corporate environmental and social responsibility and accountability. This would include actions at all levels to:

(a) Encourage industry to improve social and environmental performance through voluntary initiatives, including environmental management systems, codes of conduct, certification and public reporting on environmental and social issues, taking into account such initiatives as the International

Organization for Standardization (ISO) standards and Global Reporting Initiative guidelines on sustainability reporting, bearing in mind principle 11 of the Rio Declaration on Environment and Development;

(b) Encourage dialogue between enterprises and the communities in which they operate and other stakeholders;

(c) Encourage financial institutions to incorporate sustainable development considerations into their decision-making processes;

(d) Develop workplace-based partnerships and programmes, including training and education programmes.

18. Encourage relevant authorities at all levels to take sustainable development considerations into account in decision-making, including on national and local development planning, investment in infrastructure, business development and public procurement. This would include actions at all levels to:

(a) Provide support for the development of sustainable development strategies and programmes, including in decision-making on investment in infrastructure and business development;

(b) Continue to promote the internalization of environmental costs and the use of economic instruments, taking into account the approach that the polluter should, in principle, bear the costs of pollution, with due regard to the public interest and without distorting international trade and investment;

(c) Promote public procurement policies that encourage development and diffusion of environmentally sound goods and services;

(d) Provide capacity-building and training to assist relevant authorities with regard to the implementation of the initiatives listed in the present paragraph; and

(e) Use environmental impact assessment procedures.

* * *

19. Call upon Governments, as well as relevant regional and international organizations and other relevant stakeholders, to implement, taking into account national and regional specificities and circumstances, the recommendations and conclusions of the

Commission on Sustainable Development concerning energy for sustainable development adopted at its ninth session, including the issues and options set out below, bearing in mind that in view of the different contributions to global environmental degradation, States have common but differentiated responsibilities. This would include actions at all levels to:

(a) Take further action to mobilize the provision of financial resources, technology transfer, capacity-building and the diffusion of environmentally sound technologies according to the recommendations and conclusions of the Commission on Sustainable Development as contained in section A, paragraph 3, and section D, paragraph 30, of its decision 9/1 on energy for sustainable development;

(b) Integrate energy considerations, including energy efficiency, affordability and accessibility, into socio-economic programmes, especially into policies of major energy consuming sectors, and into the planning, operation and maintenance of long-lived energy consuming infrastructures, such as the public sector, transport, industry, agriculture, urban land use, tourism and construction sectors;

(c) Develop and disseminate alternative energy technologies with the aim of giving a greater share of the energy mix to renewable energies, improving energy efficiency and greater reliance on advanced energy technologies, including cleaner fossil fuel technologies;

(d) Combine, as appropriate, the increased use of renewable energy resources, more efficient use of energy, greater reliance on advanced energy technologies, including advanced and cleaner fossil fuel technologies, and the sustainable use of traditional energy resources, which could meet the growing need for energy services in the longer term to achieve sustainable development;

(e) Diversify energy supply by developing advanced, cleaner, more efficient, affordable and cost-effective energy technologies, including fossil fuel technologies and renewable energy technologies, hydro included, and their transfer to developing countries on concessional terms as mutually agreed. With a sense of urgency, substantially increase the global share of renewable energy sources with the objective of increasing its contribution to total energy supply, recognizing the role of national and voluntary regional targets as well as initiatives, where they exist, and ensuring that energy policies are supportive to developing countries' efforts to eradicate poverty, and regularly evaluate available data to review progress to this end;

(f) Support efforts, including through provision of financial and technical assistance to developing countries, with the involvement of the private sector, to reduce flaring and venting of gas associated with crude oil production;

(g) Develop and utilize indigenous energy sources and infrastructures for various local uses and promote rural community participation, including local Agenda-21 groups, with the support of the international community, in developing and utilizing renewable energy technologies to meet their daily energy needs to find simple and local solutions;

(h) Establish domestic programmes for energy efficiency, including, as appropriate, by accelerating the deployment of energy efficiency technologies, with the necessary support of the international community;

(i) Accelerate the development, dissemination and deployment of affordable and cleaner energy efficiency and energy conservation technologies, as well as the transfer of such technologies, in particular to developing countries, on favourable terms, including on concessional and preferential terms, as mutually agreed;

(j) Recommend that international financial institutions and other agencies' policies support developing countries, as well as countries with economies in transition, in their own efforts to establish policy and regulatory frameworks which create a level playing field between the following: renewable energy, energy efficiency, advanced energy technologies, including advanced and cleaner fossil fuel technologies, and centralized, distributed and decentralized energy systems;

(k) Promote increased research and development in the field of various energy technologies, including renewable energy, energy efficiency and advanced energy technologies, including advanced and cleaner fossil fuel technologies, both nationally and through international collaboration; strengthen national and regional research and development institutions/centres on reliable, affordable, economically viable, socially acceptable and environmentally sound energy for sustainable development;

(l) Promote networking between centres of excellence on energy for sustainable development, including regional networks, by linking competent centres on energy technologies for sustainable development that could support and promote efforts at capacity-building and technology transfer activities, particularly of developing countries, as well as serve as information clearing houses;

(m) Promote education to provide information for both men and women about available energy sources and technologies;

(n) Utilize financial instruments and mechanisms, in particular the Global Environment Facility (GEF), within its mandate, to provide financial resources to developing countries, in particular least developed countries and small island developing States, to meet their capacity needs for training, technical know-how and strengthening national institutions in reliable, affordable, economically viable, socially acceptable and environmentally sound energy, including promoting energy efficiency and conservation, renewable energy and advanced energy technologies, including advanced and cleaner fossil fuel technologies;

(o) Support efforts to improve the functioning, transparency and information about energy markets with respect to both supply and demand, with the aim of achieving greater stability and predictability and to ensure consumer access to reliable, affordable, economically viable, socially acceptable and environmentally sound energy services;

(p) Policies to reduce market distortions would promote energy systems compatible with sustainable development through the use of improved market signals and by removing market distortions, including restructuring taxation and phasing out harmful subsidies, where they exist, to reflect their environmental impacts, with such policies taking fully into account the specific needs and conditions of developing countries with the aim of minimizing the possible adverse impacts on their development;

(q) Take action, where appropriate, to phase out subsidies in this area that inhibit sustainable development, taking fully into account the specific conditions and different levels of development of individual countries and considering their adverse effect, particularly on developing countries;

(r) Governments are encouraged to improve the functioning of national energy markets in such a way that they support sustainable development, overcome market barriers and improve accessibility, taking fully into account that such policies should be decided by each country, and that its own characteristics and capabilities and level of development should be considered, especially as reflected in national sustainable development strategies, where they exist;

(s) Strengthen national and regional energy institutions or arrangements for enhancing regional and international cooperation on energy for sustainable development, in particular to assist developing countries in their domestic efforts to provide reliable, affordable, economically viable, socially acceptable and environmentally sound energy services to all sections of their populations;

(t) Countries are urged to develop and implement actions within the framework of the ninth session of the Commission on Sustainable Development, including through public-private partnerships, taking into account the different circumstances of countries, based on lessons learned by Governments, international institutions and stakeholders and including business and industry, in the field of access to energy, including renewable energy and energy-efficiency and advanced energy technologies, including advanced and cleaner fossil fuel technologies;

(u) Promote cooperation between international and regional institutions and bodies dealing with different aspects of energy for sustainable development within their existing mandate, bearing in mind paragraph 46 (h) of the Programme of Action for the Further Implementation of Agenda-21, strengthening, as appropriate, regional and national activities for the promotion of education and capacity-building regarding energy for sustainable development;

(v) Strengthen and facilitate, as appropriate, regional cooperation arrangements for promoting cross-border energy trade, including the interconnection of electricity grids and oil and natural gas pipelines;

(w) Strengthen and, where appropriate, facilitate dialogue forums among regional, national and international producers and consumers of energy.

* * *

20. Promote an integrated approach to policy-making at the national, regional and local levels for transport services and systems to promote sustainable development, including policies and planning for land use, infrastructure, public transport systems and goods delivery networks, with a view to providing safe, affordable and efficient transportation, increasing energy efficiency, reducing pollution, reducing congestion, reducing adverse health effects and limiting urban sprawl, taking into account national priorities and circumstances. This would include actions at all levels to:

(a) Implement transport strategies for sustainable development, reflecting specific regional, national and local conditions, so as to improve the affordability, efficiency and convenience of transportation, as well as improving urban air quality and health, and reduce greenhouse gas emissions, including through the development of better vehicle technologies that are more environmentally sound, affordable and socially acceptable;

(b) Promote investment and partnerships for the development of sustainable, energy efficient multi-modal transportation systems, including public mass transportation systems and better transportation systems in rural areas, with technical and financial assistance for developing countries and countries with economies in transition.

* * *

21. Prevent and minimize waste and maximize reuse, recycling and use of environmentally friendly alternative materials, with the participation of government authorities and all stakeholders, in order to minimize adverse effects on the environment and improve resource efficiency, with financial, technical and other assistance for developing countries. This would include actions at all levels to:

(a) Develop waste management systems, with highest priorities placed on waste prevention and minimization, reuse and recycling, and environmentally sound disposal facilities, including technology to recapture the energy contained in waste, and encourage smallscale waste-recycling initiatives that support urban and rural waste management and provide income-generating opportunities, with international support for developing countries;

(b) Promote waste prevention and minimization by encouraging production of reusable consumer goods and biodegradable products and developing the infrastructure required.

* * *

22. Renew the commitment, as advanced in Agenda-21, to sound management of chemicals throughout their life cycle and of hazardous wastes for sustainable development and for the protection of human health and the environment, inter alia, aiming to achieve by 2020 that chemicals are used and produced in ways that lead to the minimization of significant adverse effects on human health and the environment, using transparent science-based risk assessment procedures and science-based risk management procedures, taking into account the precautionary approach, as set out in principle 15 of the Rio Declaration on Environment and Development, and support developing countries in strengthening their capacity for the sound management of chemicals and hazardous wastes by providing technical and financial assistance. This would include actions at all levels to:

(a) Promote the ratification and implementation of relevant international instruments on chemicals and hazardous waste, including the Rotterdam Convention on Prior Informed Consent Procedures for Certain Hazardous Chemicals and Pesticides in International Trade so that it can enter into force by 2003 and the Stockholm Convention on Persistent Organic Pollutants so that it can enter into force by 2004, and encourage and improve coordination as well as supporting developing countries in their implementation;

(b) Further develop a strategic approach to international chemicals management based on the Bahia Declaration and Priorities for Action beyond 2000 of the Intergovernmental Forum on

Chemical Safety (IFCS) by 2005, and urge that the United Nations Environment Programme (UNEP), IFCS, other international organizations dealing with chemical management, and other relevant international organizations and actors closely cooperate in this regard, as appropriate;

(c) Encourage countries to implement the new globally harmonized system for the classification and labelling of chemicals as soon as possible with a view to having the system fully operational by 2008;

(d) Encourage partnerships to promote activities aimed at enhancing environmentally sound management of chemicals and hazardous wastes, implementing multilateral environmental agreements, raising awareness of issues relating to chemicals and hazardous waste, and encouraging the collection and use of additional scientific data;

(e) Promote efforts to prevent international illegal trafficking of hazardous chemicals and hazardous wastes and to prevent damage resulting from the transboundary movement and disposal of hazardous wastes in a manner consistent with obligations under relevant international instruments, such as the Basel Convention on the Control of Transboundary Movements of Hazardous Wastes and Their Disposal;

(f) Encourage development of coherent and integrated information on chemicals, such as through national pollutant release and transfer registers;

(g) Promote reduction of the risks posed by heavy metals that are harmful to human health and the environment, including through a review of relevant studies, such as the UNEP global assessment of mercury and its compounds.

PROTECTING AND MANAGING THE NATURAL RESOURCE BASE OF ECONOMIC AND SOCIAL DEVELOPMENT

23. Human activities are having an increasing impact on the integrity of ecosystems that provide essential resources and services for human well-being and economic activities. Managing the natural resources base in a sustainable and integrated manner is essential for sustainable development. In this regard, to reverse the current trend in natural resource degradation as soon as possible, it is necessary to implement strategies which should include targets adopted at the

national and, where appropriate, regional levels to protect ecosystems and to achieve integrated management of land, water and living resources, while strengthening regional, national and local capacities.

24. Launch a programme of actions, with financial and technical assistance, to achieve the millennium development goal on safe drinking water. In this respect, we agree to halve, by the year 2015, the proportion of people who are unable to reach or to afford safe drinking water as outlined in the Millennium Declaration and the proportion of people without access to basic sanitation, which would include actions at all levels to:

(a) Mobilize international and domestic financial resources at all levels, transfer technology, promote best practice and support capacity-building for water and sanitation infrastructure and services development, ensuring that such infrastructure and services meet the needs of the poor and are gender-sensitive;
(b) Facilitate access to public information and participation, including by women, at all levels, in support of policy and decision-making related to water resources management and project implementation;
(c) Promote priority action by Governments, with the support of all stakeholders, in water management and capacity-building at the national level and, where appropriate, at the regional level, and promote and provide new and additional financial resources and innovative technologies to implement chapter 18 of Agenda-21;
(d) Intensify water pollution prevention to reduce health hazards and protect ecosystems by introducing technologies for affordable sanitation and industrial and domestic wastewater treatment, by mitigating the effects of groundwater contamination, and by establishing, at the national level, monitoring systems and effective legal frameworks;
(e) Adopt prevention and protection measures to promote sustainable water use and to address water shortages.

25. Develop integrated water resources management and water efficiency plans by 2005, with support to developing countries, through actions at all levels to:

(a) Develop and implement national/regional strategies, plans and programmes with regard to integrated river basin, watershed and groundwater management, and introduce measures to improve the efficiency of water infrastructure to reduce losses and increase recycling of water;

(b) Employ the full range of policy instruments, including regulation, monitoring, voluntary measures, market and information-based tools, land-use management and cost recovery of water services, without cost recovery objectives becoming a barrier to access to safe water by poor people, and adopt an integrated water basin approach;

(c) Improve the efficient use of water resources and promote their allocation among competing uses in a way that gives priority to the satisfaction of basic human needs and balances the requirement of preserving or restoring ecosystems and their functions, in particular in fragile environments, with human domestic, industrial and agriculture needs, including safeguarding drinking water quality;

(d) Develop programmes for mitigating the effects of extreme water-related events;

(e) Support the diffusion of technology and capacity-building for non-conventional water resources and conservation technologies, to developing countries and regions facing water scarcity conditions or subject to drought and desertification, through technical and financial support and capacity-building;

(f) Support, where appropriate, efforts and programmes for energy-efficient, sustainable and cost-effective desalination of seawater, water recycling and water harvesting from coastal fogs in developing countries, through such measures as technological, technical and financial assistance and other modalities;

(g) Facilitate the establishment of public-private partnerships and other forms of partnership that give priority to the needs of the poor, within stable and transparent national regulatory frameworks provided by Governments, while respecting local conditions, involving all concerned stakeholders, and monitoring the performance and improving accountability of public institutions and private companies.

26. Support developing countries and countries with economies in transition in their efforts to monitor and assess the quantity and quality

of water resources, including through the establishment and/or further development of national monitoring networks and water resources databases and the development of relevant national indicators.

27. Improve water resource management and scientific understanding of the water cycle through cooperation in joint observation and research, and for this purpose encourage and promote knowledge-sharing and provide capacity-building and the transfer of technology, as mutually agreed, including remote-sensing and satellite technologies, particularly to developing countries and countries with economies in transition.

28. Promote effective coordination among the various international and intergovernmental bodies and processes working on water-related issues, both within the United Nations system and between the United Nations and international financial institutions, drawing on the contributions of other international institutions and civil society to inform intergovernmental decision-making; closer coordination should also be promoted to elaborate and support proposals and undertake activities related to the International Year of Freshwater 2003 and beyond.

* * *

29. Oceans, seas, islands and coastal areas form an integrated and essential component of the Earth's ecosystem and are critical for global food security and for sustaining economic prosperity and the well-being of many national economies, particularly in developing countries. Ensuring the sustainable development of the oceans requires effective coordination and cooperation, including at the global and regional levels, between relevant bodies, and actions at all levels to:

(a) Invite States to ratify or accede to and implement the United Nations Convention on the Law of the Sea, which provides the overall legal framework for ocean activities;

(b) Promote the implementation of chapter 17 of Agenda-21 which provides the programme of action for achieving the sustainable development of oceans, coastal areas and seas through its programme areas of integrated management and sustainable development of coastal areas, including exclusive economic zones; marine environmental protection; sustainable use and conservation of marine living resources; addressing critical uncertainties for the management of the marine environment and climate change; strengthening international, including regional, cooperation and coordination; and sustainable development of small islands;

(c) stablish an effective, transparent and regular inter-agency coordination mechanism on ocean and coastal issues within the United Nations system;

(d) Encourage the application by 2010 of the ecosystem approach, noting the Reykjavik Declaration on Responsible Fisheries in the Marine Ecosystem and decision 5/6 of the Conference of Parties to the Convention on Biological Diversity;

(e) Promote integrated, multidisciplinary and multisectoral coastal and ocean management at the national level, and encourage and assist coastal States in developing ocean policies and mechanisms on integrated coastal management;

(f) Strengthen regional cooperation and coordination between the relevant regional organizations and programmes, the UNEP regional seas programmes, regional fisheries management organizations and other regional science, health and development organizations;

(g) Assist developing countries in coordinating policies and programmes at the regional and subregional levels aimed at the conservation and sustainable management of fishery resources, and implement integrated coastal area management plans, including through the promotion of sustainable coastal and small-scale fishing activities and, where appropriate, the development of related infrastructure;

(h) Take note of the work of the open-ended informal consultative process established by the United Nations General Assembly in its resolution 54/33 in order to facilitate the annual review by the Assembly of developments in ocean affairs and the upcoming review of its effectiveness and utility to be held at its fifty-seventh session under the terms of the abovementioned resolution.

30. To achieve sustainable fisheries, the following actions are required at all levels:

(a) Maintain or restore stocks to levels that can produce the maximum sustainable yield with the aim of achieving these goals for depleted stocks on an urgent basis and where possible not later than 2015;

(b) Ratify or accede to and effectively implement the relevant United Nations and, where appropriate, associated regional fisheries agreements or arrangements, noting in particular the Agreement for the Implementation of the Provisions of the United Nations Convention on the Law of the Sea of 10 December 1982 relating to the Conservation and Management of Straddling Fish Stocks and Highly Migratory Fish Stocks and the 1993 Agreement to Promote Compliance with International Conservation and Management Measures by Fishing Vessels on the High Seas;

(c) Implement the 1995 Code of Conduct for Responsible Fisheries, taking note of the special requirements of developing countries as noted in its article 5, and the relevant Food and Agriculture Organization of the United Nations (FAO) international plans of action and technical guidelines;

(d) Urgently develop and implement national and, where appropriate, regional plans of action, to put into effect the FAO international plans of action, in particular the international plan of action for the management of fishing capacity by 2005 and the international plan of action to prevent, deter and eliminate illegal, unreported and unregulated fishing by 2004. Establish effective monitoring, reporting and enforcement, and control of fishing vessels, including by flag States, to further the international plan of action to prevent, deter and eliminate illegal, unreported and unregulated fishing;

(e) Encourage relevant regional fisheries management organizations and arrangements to give due consideration to the rights, duties and interests of coastal States and the special requirements of developing States when addressing the issue of the allocation of share of fishery resources for straddling stocks and highly migratory fish stocks, mindful of the provisions of the United Nations Convention on the Law of the Sea and the Agreement for the Implementation of the Provisions of the United Nations Convention on the Law of the Sea of 10 December 1982 Relating to the Conservation and Management of Straddling Fish Stocks and Highly Migratory Fish Stocks, on the high seas and within exclusive economic zones;

(f) Eliminate subsidies that contribute to illegal, unreported and unregulated fishing and to over-capacity, while completing the efforts undertaken at WTO to clarify and improve its disciplines on fisheries subsidies, taking into account the importance of this sector to developing countries;
(g) Strengthen donor coordination and partnerships between international financial institutions, bilateral agencies and other relevant stakeholders to enable developing countries, in particular the least developed countries and small island developing States and countries with economies in transition, to develop their national, regional and subregional capacities for infrastructure and integrated management and the sustainable use of fisheries;
(h) Support the sustainable development of aquaculture, including small-scale aquaculture, given its growing importance for food security and economic development.

31. In accordance with chapter 17 of Agenda-21, promote the conservation and management of the oceans through actions at all levels, giving due regard to the relevant international instruments to:

(a) Maintain the productivity and biodiversity of important and vulnerable marine and coastal areas, including in areas within and beyond national jurisdiction;
(b) Implement the work programme arising from the Jakarta Mandate on the Conservation and Sustainable Use of Marine and Coastal Biological Diversity of the Convention on Biological Diversity, including through the urgent mobilization of financial resources and technological assistance and the development of human and institutional capacity, particularly in developing countries;
(c) Develop and facilitate the use of diverse approaches and tools, including the ecosystem approach, the elimination of destructive fishing practices, the establishment of marine protected areas consistent with international law and based on scientific information, including representative networks by 2012 and time/area closures for the protection of nursery grounds and periods, proper coastal land use; and watershed planning and the integration of marine and coastal areas management into key sectors;

(d) Develop national, regional and international programmes for halting the loss of marine biodiversity, including in coral reefs and wetlands;

(e) Implement the RAMSAR Convention, including its joint work programme with the Convention on Biological Diversity, and the programme of action called for by the International Coral Reef Initiative to strengthen joint management plans and international networking for wetland ecosystems in coastal zones, including coral reefs, mangroves, seaweed beds and tidal mud flats.

32. Advance implementation of the Global Programme of Action for the Protection of the Marine Environment from Land-based Activities and the Montreal Declaration on the Protection of the Marine Environment from Land-based Activities, with particular emphasis in the period 2002-2006 on municipal wastewater, the physical alteration and destruction of habitats, and nutrients, by actions at all levels to:

(a) Facilitate partnerships, scientific research and diffusion of technical knowledge; mobilize domestic, regional and international resources; and promote human and institutional capacity-building, paying particular attention to the needs of developing countries;

(b) Strengthen the capacity of developing countries in the development of their national and regional programmes and mechanisms to mainstream the objectives of the Global Programme of Action and to manage the risks and impacts of ocean pollution;

(c) Elaborate regional programmes of action and improve the links with strategic plans for the sustainable development of coastal and marine resources, noting in particular areas which are subject to accelerated environmental changes and development pressures;

(d) Make every effort to achieve substantial progress by the next Global Programme of Action conference in 2006 to protect the marine environment from land-based activities.

33. Enhance maritime safety and protection of the marine environment from pollution by actions at all levels to:

(a) Invite States to ratify or accede to and implement the conventions and protocols and other relevant instruments of the International Maritime Organization (IMO) relating to the enhancement of maritime safety and protection of the marine environment from marine pollution and environmental damage caused by ships, including the use of toxic anti-fouling paints and urge IMO to consider stronger mechanisms to secure the implementation of IMO instruments by flag States;

(b) Accelerate the development of measures to address invasive alien species in ballast water. Urge IMO to finalize the IMO International Convention on the Control and Management of Ships' Ballast Water and Sediments.

33.(bis) Governments, taking into account their national circumstances, are encouraged, recalling paragraph 8 of resolution GC (44)/RES/17 of the General Conference of the International Atomic Energy Agency (IAEA) and taking into account the very serious potential for environment and human health impacts of radioactive wastes, to make efforts to examine and further improve measures and internationally agreed regulations regarding safety, while stressing the importance of having effective liability mechanisms in place, relevant to international maritime transportation and other transboundary movement of radioactive material, radioactive waste and spent fuel, including, inter alia, arrangements for prior notification and consultations done in accordance with relevant international instruments.

34. Improve the scientific understanding and assessment of marine and coastal ecosystems as a fundamental basis for sound decision-making, through actions at all levels to:

(a) Increase scientific and technical collaboration, including integrated assessment at the global and regional levels, including the appropriate transfer of marine science and marine technologies and techniques for the conservation and management of living and non-living marine resources and expanding ocean-observing capabilities for the timely prediction and assessment of the state of marine environment;

(b) Establish by 2004 a regular process under the United Nations for global reporting and assessment of the state of the marine environment, including socio-economic aspects, both current and foreseeable, building on existing regional assessments;

(c) Build capacity in marine science, information and management, through, inter alia, promoting the use of environmental impact assessments and environmental evaluation and reporting techniques, for projects or activities that are potentially harmful to the coastal and marine environments and their living and non-living resources;

(d) Strengthen the ability of the Intergovernmental Oceanographic Commission of the United Nations Educational, Scientific and Cultural Organization, FAO and other relevant international and regional and subregional organizations to build national and local capacity in marine science and the sustainable management of oceans and their resources.

* * *

35. An integrated, multi-hazard, inclusive approach to address vulnerability, risk assessment and disaster management, including prevention, mitigation, preparedness, response and recovery, is an essential element of a safer world in the twenty-first century. Actions are required at all levels to:

(a) Strengthen the role of the International Strategy for Disaster Reduction and encourage the international community to provide the necessary financial resources to its Trust Fund;

(b) Support the establishment of effective regional, subregional and national strategies and scientific and technical institutional support for disaster management;

(c) Strengthen the institutional capacities of countries and promote international joint observation and research, through improved surface-based monitoring and increased use of satellite data, dissemination of technical and scientific knowledge, and the provision of assistance to vulnerable countries;

(d) Reduce the risks of flooding and drought in vulnerable countries by, inter alia, promoting wetland and watershed protection and restoration, improved land-use planning, improving and applying more widely techniques and methodologies for assessing the potential adverse effects of climate change on wetlands and, as appropriate, assisting countries that are particularly vulnerable to those effects;

(e) Improve techniques and methodologies for assessing the effects of climate change, and encourage the continuing assessment of those adverse effects by the Intergovernmental Panel on Climate Change;

(f) Encourage the dissemination and use of traditional and indigenous knowledge to mitigate the impact of disasters, and promote community-based disaster management planning by local authorities, including through training activities and raising public awareness;

(g) Support the ongoing voluntary contribution of, as appropriate, non-governmental organizations, the scientific community and other partners in the management of natural disasters according to agreed, relevant guidelines;

(h) Develop and strengthen early warning systems and information networks in disaster management, consistent with the International Strategy for Disaster Reduction;

(i) Develop and strengthen capacity at all levels to collect and disseminate scientific and technical information, including the improvement of early warning systems for predicting extreme weather events, especially El Niño/La Niña, through the provision of assistance to institutions devoted to addressing such events, including the International Centre for the Study of the El Niño phenomenon;

(j) Promote cooperation for the prevention and mitigation of, preparedness for, response to and recovery from major technological and other disasters with an adverse impact on the environment in order to enhance the capabilities of affected countries to cope with such situations.

* * *

36. Change in the Earth's climate and its adverse effects are a common concern of humankind. We remain deeply concerned that all countries, particularly developing countries including the least developed countries and small island developing States, face increased risks of negative impacts of climate change and recognize that, in this context, the problems of poverty, land degradation, access to water and food and human health remain at the centre of global attention. The United Nations Framework Convention on Climate Change is the key instrument for addressing climate change, a global concern, and

we reaffirm our commitment to achieving its ultimate objective of stabilization of greenhouse gas concentrations in the atmosphere at a level that would prevent dangerous anthropogenic interference with the climate system, within a time frame sufficient to allow ecosystems to adapt naturally to climate change, to ensure that food production is not threatened and to enable economic development to proceed in a sustainable manner, in accordance with our common but differentiated responsibilities and respective capabilities. Recalling the United Nations Millennium Declaration, in which heads of State and Government resolved to make every effort to ensure the entry into force of the Kyoto Protocol to the United Nations Framework Convention on Climate Change, preferably by the tenth anniversary of the United Nations Conference on Environment and Development in 2002, and to embark on the required reduction of emissions of greenhouse gases, States that have ratified the Kyoto Protocol strongly urge States that have not already done so to ratify the Kyoto Protocol in a timely manner. Actions at all levels are required to:

(a) Meet all the commitments and obligations under the UNFCCC;
(b) Work cooperatively towards achieving the objectives of the UNFCCC;
(c) Provide technical and financial assistance and capacity building to developing countries and countries with economies in transition in accordance with commitments under the UNFCCC, including the Marrakech accords;
(d) Build and enhance scientific and technological capabilities, inter alia through continuing support to the IPCC for the exchange of scientific data and information especially in developing countries;
(e) Develop and transfer technological solutions;
(f) Develop and disseminate innovative technologies in respect of key sectors of development, particularly energy, and of investment in this regard, including through private sector involvement, market-oriented approaches, as well as supportive public policies and international cooperation;
(g) Promote the systematic observation of the Earth's atmosphere, land and oceans by improving monitoring stations, increasing the use of satellites, and appropriate integration of these observations to produce high-quality data that could be disseminated for the use of all countries, in particular

developing countries;

(h) Enhance the implementation of national, regional and international strategies to monitor the Earth's atmosphere, land and oceans including, as appropriate, strategies for integrated global observations, inter alia with the cooperation of relevant international organisations, especially the United Nations specialized agencies in cooperation with the UNFCCC;

(i) Support initiatives to assess the consequences of climate change, such as the Arctic Council initiative, including the environmental, economic and social impacts on local and indigenous communities.

37. Enhance cooperation at the international, regional and national levels to reduce air pollution, including transboundary air pollution, acid deposition and ozone depletion bearing in mind the Rio principles, including, inter alia, the principle that, in view of the different contributions to global environmental degradation, States have common but differentiated responsibilities, with actions at all levels to:

(a) Strengthen capacities of developing countries and countries with economies in transition to measure, reduce and assess the impacts of air pollution, including health impacts, and provide financial and technical support for these activities;

(b) Facilitate implementation of the Montreal Protocol on Substances that Deplete the Ozone Layer by ensuring adequate replenishment of its fund by 2003/2005;

(c) Further support the effective regime for the protection of the ozone layer established in the Vienna Convention for the Protection of the Ozone Layer and the Montreal Protocol, including its compliance mechanism;

(d) Improve access by developing countries to affordable, accessible, cost-effective, safe and environmentally sound alternatives to ozone-depleting substances by 2010, and assist them in complying with the phase-out schedule under the Montreal Protocol, bearing in mind that ozone depletion and climate change are scientifically and technically interrelated;

(e) Take measures to address illegal traffic in ozone-depleting substances.

* * *

38. Agriculture plays a crucial role in addressing the needs of a growing global population, and is inextricably linked to poverty eradication, especially in developing countries. Enhancing the role of women at all levels and in all aspects of rural development, agriculture, nutrition and food security is imperative. Sustainable agriculture and rural development are essential to the implementation of an integrated approach to increasing food production and enhancing food security and food safety in an environmentally sustainable way. This would include actions at all levels to:

(a) Achieve the Millennium Declaration target to halve by the year 2015 the proportion of the world's people who suffer from hunger and realize the right to a standard of living adequate for the health and well-being of themselves and their families, including food, including by promoting food security and fighting hunger in combination with measures which address poverty, consistent with the outcome of the World Food Summit and, for States Parties, with their obligations under article 11 of the International Covenant on Economic, Social and Cultural Rights;

(b) Develop and implement integrated land management and water-use plans that are based on sustainable use of renewable resources and on integrated assessments of socioeconomic and environmental potentials, and strengthen the capacity of Governments, local authorities and communities to monitor and manage the quantity and quality of land and water resources;

(c) Increase understanding of the sustainable use, protection and management of water resources to advance long-term sustainability of freshwater, coastal and marine environments;

(d) Promote programmes to enhance in a sustainable manner the productivity of land and the efficient use of water resources in agriculture, forestry, wetlands, artisanal fisheries and aquaculture, especially through indigenous and local community-based approaches;

(e) Support the efforts of developing countries to protect oases from silt, land degradation and increasing salinity by providing appropriate technical and financial assistance;

(f) Enhance the participation of women in all aspects and at all levels relating to sustainable agriculture and food security;

(g) Integrate existing information systems on land-use practices by strengthening national research and extension services and farmer organizations to trigger farmer-to-farmer exchange on good practices, such as those related to environmentally sound, low-cost technologies, with the assistance of relevant international organizations;

(h) Enact, as appropriate, measures that protect indigenous resource management systems and support the contribution of all appropriate stakeholders, men and women alike, in rural planning and development;

(i) Adopt policies and implement laws that guarantee well defined and enforceable land and water use rights, and promote legal security of tenure, recognizing the existence of different national laws and/or systems of land access and tenure, and provide technical and financial assistance to developing countries as well as countries with economies in transition that are undertaking land tenure reform in order to enhance sustainable livelihoods;

(j) Reverse the declining trend in public sector finance for sustainable agriculture, provide appropriate technical and financial assistance, and promote private sector investment and support efforts in developing countries and countries with economies in transition to strengthen agricultural research and natural resource management capacity and dissemination of research results to the farming communities;

(k) Employ market-based incentives for agricultural enterprises and farmers to monitor and manage water use and quality, inter alia, by applying such methods as small-scale irrigation and wastewater recycling and reuse;

(l) Enhance access to existing markets and develop new markets for value-added agricultural products;

(m) Increase brown-field redevelopment in developed countries and countries with economies in transition, with appropriate technical assistance where contamination is a serious problem;

(n) Enhance international cooperation to combat the illicit cultivation of narcotic plants, taking into account their negative social, economic and environmental impacts;

(o) Promote programmes for the environmentally sound, effective and efficient use of soil fertility improvement practices and agricultural pest control;

(p) Strengthen and improve coordination of existing initiatives to enhance sustainable agricultural production and food security;
(q) Invite countries that have not done so to ratify the International Treaty on Plant Genetic Resources for Food and Agriculture;
(r) Promote the conservation, and sustainable use and management of traditional and indigenous agricultural systems and strengthen indigenous models of agricultural production.

* * *

39. Strengthen the implementation of the United Nations Convention to Combat Desertification in Those Countries Experiencing Serious Drought and/or Desertification, particularly in Africa, to address causes of desertification and land degradation in order to maintain and restore land, and to address poverty resulting from land degradation. This would include actions at all levels to:

(a) Mobilize adequate and predictable financial resources, transfer of technologies and capacity-building at all levels;
(b) Formulate national action programmes to ensure timely and effective implementation of the Convention and its related projects, with the support of the international community, including through decentralized projects at the local level;
(c) Encourage the United Nations Framework Convention on Climate Change, the Convention on Biological Diversity and the Convention to Combat Desertification to continue exploring and enhancing synergies, with due regard to their respective mandates, in the elaboration and implementation of plans and strategies under the respective Conventions;
(d) Integrate measures to prevent and combat desertification as well as to mitigate the effects of drought through relevant policies and programmes, such as land, water and forest management, agriculture, rural development, early warning systems, environment, energy, natural resources, health and education, and poverty eradication and sustainable development strategies;
(e) Provide affordable local access to information to improve monitoring and early warning related to desertification and drought;
(f) Call on the Second Assembly of the Global Environment

Facility (GEF) to take action on the recommendations of the GEF Council concerning the designation of land degradation (desertification and deforestation) as a focal area of GEF as a means of GEF support for the successful implementation of the Convention to Combat Desertification; and consequently, consider making GEF a financial mechanism of the Convention, taking into account the prerogatives and decisions of the Conference of the Parties to the Convention, while recognizing the complementary roles of GEF and the Global Mechanism of the Convention in providing and mobilizing resources for the elaboration and implementation of action programmes;

(g) Improve the sustainability of grassland resources through strengthening management and law enforcement and providing financial and technical support by the international community to developing countries.

* * *

40. Mountain ecosystems support particular livelihoods, and include significant watershed resources, biological diversity and unique flora and fauna. Many are particularly fragile and vulnerable to the adverse effects of climate change and need specific protection. Actions at all levels are required to:

(a) Develop and promote programmes, policies and approaches that integrate environmental, economic and social components of sustainable mountain development and strengthen international cooperation for its positive impacts on poverty eradication programmes, especially in developing countries;

(b) Implement programmes to address, where appropriate, deforestation, erosion, land degradation, loss of biodiversity, disruption of water flows and retreat of glaciers;

(c) Develop and implement, where appropriate, gender-sensitive policies and programmes, including public and private investments that help eliminate inequities facing mountain communities;

(d) Implement programmes to promote diversification and traditional mountain economies, sustainable livelihoods and small-scale production systems, including specific training programmes and better access to national and international markets, communications and transport planning, taking into account the particular sensitivity of mountains;

(e) Promote full participation and involvement of mountain communities in decisions that affect them and integrate indigenous knowledge, heritage and values in all development initiatives;

(f) Mobilize national and international support for applied research and capacitybuilding, provide financial and technical assistance for the effective implementation of sustainable development of mountain ecosystems in developing countries and countries with economies in transition, and address the poverty among people living in mountains through concrete plans, projects and programmes, with sufficient support from all stakeholders, taking into account the spirit of the International Year of the Mountain 2002.

* * *

41. Promote sustainable tourism development, including non-consumptive and eco-tourism, taking into account the spirit of the International Year of Eco-tourism 2002, the United Nations Year for Cultural Heritage in 2002, the World Eco-tourism Summit 2002 and its Quebec Declaration, and the Global Code of Ethics for Tourism as adopted by the World Tourism Organization in order to increase the benefits from tourism resources for the population in host communities while maintaining the cultural and environmental integrity of the host communities and enhancing the protection of ecologically sensitive areas and natural heritages. Promote sustainable tourism development and capacity-building in order to contribute to the strengthening of rural and local communities. This would include actions at all levels to:

(a) Enhance international cooperation, foreign direct investment and partnerships with both private and public sectors, at all levels;

(b) Develop programmes, including education and training programmes, that encourage people to participate in eco-tourism, enable indigenous and local communities to develop and benefit from eco-tourism, and enhance stakeholder cooperation in tourism development and heritage preservation, in order to improve the protection of the environment, natural resources and cultural heritage;

(c) Provide technical assistance to developing countries and countries with economies in transition to support sustainable tourism business development and investment and tourism awareness programmes, to improve domestic tourism, and to stimulate entrepreneurial development;

(d) Assist host communities in managing visits to their tourism attractions for their maximum benefit, while ensuring the least negative impacts on and risks for their traditions, culture and environment, with the support of the World Tourism Organization and other relevant organizations;

(e) Promote the diversification of economic activities, including through the facilitation of access to markets and commercial information, and participation of emerging local enterprises, especially small and medium-sized enterprises.

* * *

42. Biodiversity, which plays a critical role in overall sustainable development and poverty eradication, is essential to our planet, human well-being and to the livelihood and cultural integrity of people. However, biodiversity is currently being lost at unprecedented rates due to human activities; this trend can only be reversed if the local people benefit from the conservation and sustainable use of biological diversity, in particular in countries of origin of genetic resources, in accordance with article 15 of the Convention on Biological Diversity. The Convention is the key instrument for the conservation and sustainable use of biological diversity and the fair and equitable sharing of benefits arising from use of genetic resources. A more efficient and coherent implementation of the three objectives of the Convention and the achievement by 2010 of a significant reduction in the current rate of loss of biological diversity will require the provision of new and additional financial and technical resources to developing countries, and includes actions at all levels to:

(a) Integrate the objectives of the Convention into global, regional and national sectoral and cross-sectoral programmes and policies, in particular in the programmes and policies of the economic sectors of countries and international financial institutions;

(b) Promote the ongoing work under the Convention on the sustainable use on biological diversity, including on sustainable tourism, as a cross-cutting issue relevant to different ecosystems, sectors and thematic areas;

(c) Encourage effective synergies between the Convention and other multilateral environmental agreements, inter alia, through the development of joint plans and programmes, with due regard to their respective mandates, regarding common responsibilities and concerns;

(d) Implement the Convention and its provisions, including active follow-up of its work programmes and decisions through national, regional and global action programmes, in particular the national biodiversity strategies and action plans, and strengthen their integration into relevant cross-sectoral strategies, programmes and policies, including those related to sustainable development and poverty eradication, including initiatives which promote community-based sustainable use of biological diversity;

(e) Promote the wide implementation and further development of the ecosystem approach, as being elaborated in the ongoing work of the Convention;

(f) Promote concrete international support and partnership for the conservation and sustainable use of biodiversity, including in ecosystems, at World Heritage sites and for the protection of endangered species, in particular through the appropriate channelling of financial resources and technology to developing countries and countries with economies in transition;

(g) To effectively conserve and sustainably use biodiversity, promote and support initiatives for hot spot areas and other areas essential for biodiversity and promote the development of national and regional ecological networks and corridors;

(h) Provide financial and technical support to developing countries, including capacitybuilding, in order to enhance indigenous and community-based biodiversity conservation efforts;

(i) Strengthen national, regional and international efforts to control invasive alien species, which are one of the main causes of biodiversity loss, and encourage the development of effective work programme on invasive alien species at all levels;

(j) Subject to national legislation, recognize the rights of local and indigenous communities who are holders of traditional knowledge, innovations and practices, and, with the approval and involvement of the holders of such knowledge, innovations and practices, develop and implement benefit-sharing mechanisms on mutually agreed terms for the use of such knowledge, innovations and practices;

(k) Encourage and enable all stakeholders to contribute to the implementation of the objectives of the Convention, and in particular recognize the specific role of youth, women and indigenous and local communities in conserving and using biodiversity in a sustainable way;

(l) Promote the effective participation of indigenous and local communities in decision and policy-making concerning the use of their traditional knowledge;

(m) Encourage technical and financial support to developing countries and countries with economies in transition in their efforts to develop and implement, as appropriate, inter alia, national sui generis systems and traditional systems according to national priorities and legislation, with a view to conserving and the sustainable use of biodiversity;

(n) Promote the wide implementation of and continued work on the Bonn Guidelines on Access to Genetic Resources and Fair and Equitable Sharing of Benefits arising out of their Utilization of the Convention, as an input to assist Parties to the Convention when developing and drafting legislative, administrative or policy measures on access and benefit-sharing, and contract and other arrangements under mutually agreed terms for access and benefit-sharing;

(o) Negotiate within the framework of the Convention on Biological Diversity, bearing in mind the Bonn Guidelines, an international regime to promote and safeguard the fair and equitable sharing of benefits arising out of the utilization of genetic resources;

(p) Encourage successful conclusion of existing processes under the World Intellectual Property Organization Inter-governmental Committee on Intellectual Property and Genetic

Resources, Traditional Knowledge and Folklore, and in the ad hoc open-ended working group on article 8 (j) and related provisions of the Convention;

(q) Promote practicable measures for access to the results and benefits arising from biotechnologies based upon genetic resources, in accordance with articles 15 and 19 of the Convention, including through enhanced scientific and technical cooperation on biotechnology and biosafety, including the exchange of experts, training human resources and developing research-oriented institutional capacities;

(r) With a view to enhancing synergy and mutual supportiveness, taking into account the decisions under the relevant agreements, promote the discussions, without prejudging their outcome, with regard to the relationships between the Convention and agreements related to international trade and intellectual property rights, as outlined in the Doha Ministerial Declaration;

(s) Promote the implementation of the programme of work of the Global Taxonomy Initiative;

(t) Invite all States which have not already done so to ratify the Convention, the Cartagena Protocol on Biosafety and other biodiversity-related agreements, and invite those that have done so, to promote their effective implementation at the national, regional and international levels and to support developing countries and countries with economies in transition technically and financially in this regard.

* * *

43. Forests and trees cover nearly one third of the Earth's surface. Sustainable forest management of both natural and planted forests and for timber and non-timber products is essential to achieving sustainable development and is a critical means to eradicate poverty, significantly reduce deforestation and halt the loss of forest biodiversity and land and resource degradation, and improve food security and access to safe drinking water and affordable energy; highlights the multiple benefits of both natural and planted forests and trees; and contributes to the well-being of the planet and humanity. Achievement of sustainable forest management, nationally and globally, including through partnerships among interested Governments and stakeholders,

including the private sector, indigenous and local communities and nongovernmental organizations, is an essential goal of sustainable development. This would include actions at all levels to:

(a) Enhance political commitment to achieve sustainable forest management by endorsing it as a priority on the international political agenda, taking full account of the linkages between the forest sector and other sectors through integrated approaches;
(b) Support the United Nations Forum on Forests, with the assistance of the Collaborative Partnership on Forests, as key intergovernmental mechanisms to facilitate and coordinate the implementation of sustainable forest management at the national, regional and global levels, thus contributing, inter alia, to the conservation and sustainable use of forest biodiversity;
(c) Take immediate action on domestic forest law enforcement and illegal international trade in forest products, including in forest biological resources, with the support of the international community, and provide human and institutional capacity-building related to the enforcement of national legislation in those areas;
(d) Take immediate action at the national and international levels to promote and facilitate the means to achieve sustainable timber harvesting, and to facilitate the provision of financial resources and the transfer and development of environmentally sound technologies, and thereby address unsustainable timber-harvesting practices;
(e) Develop and implement initiatives to address the needs of those parts of the world that currently suffer from poverty and the highest rates of deforestation and where international cooperation would be welcomed by affected Governments;
(f) Create and strengthen partnerships and international cooperation to facilitate the provision of increased financial resources, the transfer of environmentally sound technologies, trade, capacity-building, forest law enforcement and governance at all levels, and integrated land and resource management to implement sustainable forest management, including the Intergovernmental Panel on Forests (IPF)/ Intergovernmental Forum on Forests (IFF) proposals for

action;

(g) Accelerate implementation of the IPF/IFF proposals for action by countries and by the Collaborative Partnership on Forests, and intensify efforts on reporting to the United Nations Forum on Forests to contribute to an assessment of progress in 2005;

(h) Recognize and support indigenous and community-based forest management systems to ensure their full and effective participation in sustainable forest management;

(i) Implement the Convention on Biological Diversity's expanded action-oriented work programme on all types of forest biological diversity, in close cooperation with the Forum, Partnership members and other forest-related processes and conventions, with the involvement of all relevant stakeholders.

* * *

44. Mining, minerals and metals are important to the economic and social development of many countries. Minerals are essential for modern living. Enhancing the contribution of mining, minerals and metals to sustainable development includes actions at all levels to:

(a) Support efforts to address the environmental, economic, health and social impacts and benefits of mining, minerals and metals throughout their life cycle, including workers' health and safety, and use a range of partnerships, furthering existing activities at the national and international levels, among interested Governments, intergovernmental organizations, mining companies and workers, and other stakeholders, to promote transparency and accountability for sustainable mining and minerals development;

(b) Enhance the participation of stakeholders, including local and indigenous communities and women, to play an active role in minerals, metals and mining development throughout the life cycles of mining operations, including after closure for rehabilitation purposes, in accordance with national regulations and taking into account significant transboundary impacts;

(c) Foster sustainable mining practices through the provision of financial, technical and capacity-building support to developing countries and countries with economies in transition for the mining and processing of minerals, including small-scale mining, and, where possible and appropriate, improve value-added processing, upgrade scientific and technological information, and reclaim and rehabilitate degraded sites.

SUSTAINABLE DEVELOPMENT IN A GLOBALIZING WORLD*

45. Globalization offers opportunities and challenges for sustainable development. We recognize that globalization and interdependence are offering new opportunities to trade, investment and capital flows and advances in technology, including information technology, for the growth of the world economy, development and the improvement of living standards around the world. At the same time, there remain serious challenges, including serious financial crises, insecurity, poverty, exclusion and inequality within and among societies. The developing countries and countries with economies in transition face special difficulties in responding to those challenges and opportunities. Globalization should be fully inclusive and equitable, and there is a strong need for policies and measures at the national and international levels, formulated and implemented with the full and effective participation of developing countries and countries with economies in transition, to help them to respond effectively to those challenges and opportunities. This will require urgent action at all levels to:

(a) Continue to promote open, equitable, rules-based, predictable and non-discriminatory multilateral trading and financial systems that benefit all countries in the pursuit of sustainable development. Support the successful completion of the work programme contained in the Doha Ministerial Declaration and the implementation of the Monterrey Consensus. Welcome the decision contained in the Doha Ministerial Declaration to place the needs and interests of developing countries at the

---

* Because the structure of chapter V was changed in the course of negotiations so that its paragraph numbering no longer corresponds to the earlier version contained in document A/CONF.199/L.1, Latin ordinal numbers (bis, ter etc.) are used after paragraph 45 until the end of the chapter in order not to disrupt the paragraph numbering of subsequent chapters which were not so changed.

heart of the work programme of the Declaration, including through enhanced market access for products of interest to developing countries;

(b) Encourage ongoing efforts by international financial and trade institutions to ensure that decision-making processes and institutional structures are open and transparent;

(c) Enhance the capacities of developing countries, including the least developed countries, landlocked developing countries and small island developing States, to benefit from liberalized trade opportunities, through international cooperation and measures aimed at improving productivity, commodity diversification and competitiveness, community-based entrepreneurial capacity, and transportation and communication infrastructure development;

(d) Support the International Labour Organization and encourage its ongoing work on the social dimension of globalization, as stated in paragraph 64 of the Monterrey Consensus;

(e) Enhance the delivery of coordinated, effective and targeted trade-related technical assistance and capacity-building programmes, including to take advantage of existing and future market access opportunities, and to examine the relationship between trade, environment and development.

45.bis Implement the outcomes of the Doha Ministerial Conference by WTO members, further strengthen trade-related technical assistance and capacity-building, and ensure the meaningful, effective and full participation of developing countries in multilateral trade negotiations by placing their needs and interests at the heart of the WTO work programme.

45. ter Actively promote corporate responsibility and accountability, based on the Rio Principles, including through the full development and effective implementation of intergovernmental agreements and measures, international initiatives and public-private partnerships, and appropriate national regulations, and support continuous improvement in corporate practices in all countries.

45.quater Strengthen the capacities of developing countries to encourage public/private initiatives that enhance the ease of access, accuracy, timeliness and coverage of information on countries and financial markets. Multilateral and regional financial institutions could provide further assistance for these purposes.

45.quinquies Strengthen regional trade and cooperation agreements, consistent with the multilateral trading system, among developed and developing countries and countries with economies in transition, as well as among developing countries, with the support of international finance institutions and regional development banks, as appropriate, with a view to achieving the objectives of sustainable development.

45.sexties Assist developing countries and countries with economies in transition in narrowing the digital divide, creating digital opportunities and harnessing the potential of information and communication technologies for development, through technology transfer on mutually agreed terms and the provision of financial and technical support, and in this context support the World Summit on the Information Society.

HEALTH AND SUSTAINABLE DEVELOPMENT

46. The Rio Declaration on Environment and Development states that human beings are at the centre of concerns for sustainable development, and that they are entitled to a healthy and productive life, in harmony with nature. The goals of sustainable development can only be achieved in the absence of a high prevalence of debilitating diseases, while obtaining health gains for the whole population requires poverty eradication. There is an urgent need to address the causes of ill health, including environmental causes, and their impact on development, with particular emphasis on women and children, as well as vulnerable groups of society, such as people with disabilities, elderly persons and indigenous people.

47. Strengthen the capacity of health-care systems to deliver basic health services to all, in an efficient, accessible and affordable manner aimed at preventing, controlling and treating diseases, and to reduce environmental health threats, in conformity with human rights and fundamental freedoms and consistent with national laws and cultural and religious values, taking into account the reports of relevant United Nations conferences and summits and of special sessions of the General Assembly. This would include actions at all levels to:

(a) Integrate the health concerns, including those of the most vulnerable populations, into strategies, policies and programmes for poverty eradication and sustainable development;

(b) Promote equitable and improved access to affordable and efficient health-care services, including prevention, at all levels of the health system, essential and safe drugs at affordable prices, immunization services and safe vaccines, and medical technology;

(c) Provide technical and financial assistance to developing countries and countries with economies in transition to implement the Health for All Strategy, including health information systems and integrated databases on development hazards;

(d) Improve the development and management of human resources in health-care services;

(e) Promote and develop partnerships to enhance health education with the objective of achieving improved health literacy on a global basis by 2010, with the involvement of United Nations agencies, as appropriate;

(f) Develop programmes and initiatives to reduce, by the year 2015, mortality rates for infants and children under 5 by two thirds, and maternal mortality rates by three quarters, of the prevailing rate in 2000, and reduce disparities between and within developed and developing countries as quickly as possible, with particular attention to eliminating the pattern of disproportionate and preventable mortality among girl infants and children;

(g) Target research efforts and apply research results to priority public health issues, in particular those affecting susceptible and vulnerable populations, through the development of new vaccines, reducing exposures to health risks, building on equal access to health-care services, education, training and medical treatment and technology, and addressing the secondary effects of poor health;

(h) Promote thc preservation, development and use of effective traditional medicine knowledge and practices, where appropriate, in combination with modern medicine, recognizing indigenous and local communities as custodians of traditional knowledge and practices, while promoting effective protection of traditional knowledge, as appropriate, consistent with international law;

(i) Ensure equal access of women to health-care services, giving particular attention to maternal and emergency obstetric care;

(j) Address effectively, for all individuals of appropriate age, the promotion of their healthy lives, including their reproductive and sexual health, consistent with the commitments and outcomes of recent United Nations conferences and summits, including the World Summit for Children, the United Nations Conference on Environment and Development, the International Conference of Population and Development, the World Summit for Social Development and the Fourth World Conference on Women, and their respective reviews and reports;

(k) Launch international capacity-building initiatives, as appropriate, that assess health and environment linkages and use the knowledge gained to create more effective national and regional policy responses to environmental threats to human health;

(l) Transfer and disseminate, on mutually agreed terms, including through public-private multisector partnerships, technologies for safe water, sanitation and waste management for rural and urban areas in developing countries and countries with economies in transition, with international financial support, taking into account country-specific conditions and gender equality including specific technology needs of women;

(m) Strengthen and promote ILO and World Health Organization (WHO) programmes to reduce occupational deaths, injuries and illnesses, and link occupational health with public health promotion as a means of promoting public health and education;

(n) Improve availability and access for all to sufficient, safe, culturally acceptable and nutritionally adequate food, increase consumer health protection, address issues of micronutrient deficiency, and implement existing internationally agreed commitments and relevant standards and guidelines;

(o) Develop or strengthen, where applicable, preventive, promotive and curative programmes to address non-communicable diseases and conditions, such as cardiovascular diseases, cancer, diabetes, chronic respiratory diseases, injuries, violence and mental health disorders and associated risk factors, including alcohol, tobacco, unhealthy diets and lack of physical activity.

48. Implement, within the agreed time frames, all commitments agreed in the Declaration of Commitment on HIV/AIDS adopted by the General Assembly at its twenty-sixth special session, emphasizing in particular the reduction of HIV prevalence among young men and women aged 15-24 by 25 per cent in the most affected countries by 2005 and globally by 2010, as well as combat malaria, tuberculosis and other diseases by, inter alia:

(a) Implementing national preventive and treatment strategies, regional and international cooperation measures, and the development of international initiatives to provide special assistance to children orphaned by HIV/AIDS;
(b) Fulfilling commitments for the provision of sufficient resources to support the Global Fund to Fight AIDS, Tuberculosis and Malaria, while promoting access to the Fund by countries most in need;
(c) Protecting the health of workers and promoting occupational safety, by, inter alia, taking into account, as appropriate the voluntary ILO code of practice on HIV/AIDS and the world of work, to improve conditions of the workplace;
(d) Mobilizing adequate public and encouraging private financial resources for research and development on diseases of the poor, such as HIV/AIDS, malaria, tuberculosis, directed at biomedical and health research, as well as new vaccine and drug development.

49. Reduce respiratory diseases and other health impacts resulting from air pollution, with particular attention to women and children, by:

(a) Strengthening regional and national programmes, including through public-private partnerships, with technical and financial assistance to developing countries;
(b) Supporting the phasing out of lead in gasoline;
(c) Strengthening and supporting efforts for the reduction of emissions, through the use of cleaner fuels and modern pollution control techniques;
(d) Assisting developing countries in providing affordable energy to rural communities, particularly to reduce dependence on traditional fuel sources for cooking and heating, which affect

the health of women and children.

50. Phase out lead in lead-based paints and other sources of human exposure, work to prevent, in particular, children's exposure to lead, and strengthen monitoring and surveillance efforts and the treatment of lead poisoning.

[*Paragraph 51 is deleted*]

SUSTAINABLE DEVELOPMENT OF SMALL ISLAND DEVELOPING STATES

52. Small island developing States are a special case both for environment and development. Although they continue to take the lead in the path towards sustainable development in their countries, they are increasingly constrained by the interplay of adverse factors clearly underlined in Agenda-21, the Programme of Action for the Sustainable Development of Small Island Developing States and the decisions adopted at the twenty-second special session of the General Assembly. This would include actions at all levels to:

(a) Accelerate national and regional implementation of the Programme of Action, with adequate financial resources, including through GEF focal areas, transfer of environmentally sound technologies and assistance for capacity-building from the international community;
(b) Further implement sustainable fisheries management and improve financial returns from fisheries by supporting and strengthening relevant regional fisheries management organizations, as appropriate, such as the recently established Caribbean Regional Fisheries Mechanism and such agreements as the Convention on the Conservation and Management of Highly Migratory Fish Stocks in the Western and Central Pacific Ocean;
(c) Assist small island developing States, including through the elaboration of specific initiatives, in delimiting and managing in a sustainable manner their coastal areas and exclusive economic zones and the continental shelf (including, where appropriate, the continental shelf areas beyond 200 miles from coastal baselines), as well as relevant regional management

initiatives within the context of the United Nations Convention on the Law of the Sea and the UNEP regional seas programmes;

(d) Provide support, including for capacity-building, for the development and further implementation of:

 (i) Small island developing States-specific components within programmes of work on marine and coastal biological diversity;
 (ii) Freshwater programmes for small island developing States, including through the GEF focal areas;

(e) Effectively reduce, prevent and control waste and pollution and their health-related impacts by undertaking by 2004 initiatives aimed at implementing the Global Programme of Action for the Protection of the Marine Environment from Land-based Activities in small island developing States;
(f) Work to ensure that, in the ongoing negotiations and elaboration of the WTO work programme on trade in small economies, due account is taken of small island developing States, which have severe structural handicaps in integrating into the global economy, within the context of the Doha development agenda;
(g) Develop community-based initiatives on sustainable tourism by 2004, and build the capacities necessary to diversify tourism products, while protecting culture and traditions, and effectively conserving and managing natural resources;
(h) Extend assistance to small island developing States in support of local communities and appropriate national and regional organizations of small island developing States for comprehensive hazard and risk management, disaster prevention, mitigation and preparedness, and help relieve the consequences of disasters, extreme weather events and other emergencies;
(i) Support the finalization and subsequent early operationalization, on agreed terms, of economic, social and environmental vulnerability indices and related indicators as tools for the achievement of the sustainable development of the small island developing States;
(j) Assist small island developing States in mobilizing adequate

resources and partnerships for their adaptation needs relating to the adverse effects of climate change, sea level rise and climate variability, consistent with commitments under the United Nations Framework Convention on Climate Changes, where applicable;

(k) Support efforts by small island developing States to build capacities and institutional arrangements to implement intellectual property regimes;

53. Support the availability of adequate, affordable and environmentally sound energy services for the sustainable development of small island developing States by, inter alia:

(a) Strengthening ongoing and supporting new efforts on energy supply and services, by 2004, including through the United Nations system and partnership initiatives;

(b) Developing and promoting efficient use of sources of energy, including indigenous sources and renewable energy, and building the capacities of small island developing States for training, technical know-how and strengthening national institutions in the area of energy management;

54. Provide support to SIDS to develop capacity and strengthen:

(a) Health-care services for promoting equitable access to health care;

(b) Health systems for making available necessary drugs and technology in a sustainable and affordable manner to fight and control communicable and non-communicable diseases, in particular HIV/AIDS, tuberculosis, diabetes, malaria and dengue fever;

(c) Efforts to reduce and manage waste and pollution and building capacity for maintaining and managing systems to deliver water and sanitation services, in both rural and urban areas;

(d) Efforts to implement initiatives aimed at poverty eradication, which have been outlined in section II of the present document.

55. Undertake a full and comprehensive review of the implementation of the Barbados Programme of Action for the Sustainable Development of Small Island Developing States in 2004, in accordance

with the provisions set forth in General Assembly resolution S-22/2, and in this context requests the General Assembly at its fifty-seventh session to consider convening an international meeting for the sustainable development of small island developing States.

SUSTAINABLE DEVELOPMENT FOR AFRICA

56. Since the United Nations Conference on Environment and Development, sustainable development has remained elusive for many African countries. Poverty remains a major challenge and most countries on the continent have not benefited fully from the opportunities of globalization, further exacerbating the continent's marginalization. Africa's efforts to achieve sustainable development have been hindered by conflicts, insufficient investment, limited market access opportunities and supply side constraints, unsustainable debt burdens, historically declining ODA levels and the impact of HIV/AIDS. The World Summit on Sustainable Development should reinvigorate the commitment of the international community to address these special challenges and give effect to a new vision based on concrete actions for the implementation of Agenda-21 in Africa. The New Partnership for Africa's Development (NEPAD) is a commitment by African leaders to the people of Africa. It recognizes that partnerships among African countries themselves and between them and with the international community are key elements of a shared and common vision to eradicate poverty, and furthermore it aims to place their countries, both individually and collectively, on a path of sustained economic growth and sustainable development, while participating actively in the world economy and body politic. It provides a framework for sustainable development on the continent to be shared by all Africa's people. The international community welcomes NEPAD and pledges its support to the implementation of this vision, including through utilization of the benefits of South-South cooperation supported, inter alia, by the Tokyo International Conference on African Development. It also pledges support for other existing development frameworks that are owned and driven nationally by African countries and that embody poverty reduction strategies, including poverty reduction strategy papers. Achieving sustainable development includes actions at all levels to:

(a) Create an enabling environment at the regional, subregional, national and local levels in order to achieve sustained economic growth and sustainable development and support African efforts for peace, stability and security, the resolution and prevention of conflicts, democracy, good governance, respect for human rights and fundamental freedoms, including the right to development and gender equality;

(b) Support the implementation of the vision of NEPAD and other established regional and subregional efforts, including through financing, technical cooperation and institutional cooperation, and human and institutional capacity-building at the regional, subregional and national levels, consistent with national policies, programmes and nationally owned and led strategies for poverty reduction and sustainable development, such as, where applicable, poverty reduction strategy papers;

(c) Promote technology development, transfer and diffusion to Africa and further develop technology and knowledge available in African centres of excellence;

(d) Support African countries to develop effective science and technology institutions and research activities capable of developing and adapting to world class technologies;

(e) Support the development of national programmes and strategies to promote education within the context of nationally owned and led strategies for poverty reduction, and strengthen research institutions in education in order to increase the capacity to fully support the achievement of internationally agreed development goals related to education, including those contained in the Millennium Declaration on ensuring that, by 2015, children everywhere, boys and girls alike, will be able to complete a full course of primary schooling, and that girls and boys will have equal access to all levels of education relevant to national needs;

(f) Enhance the industrial productivity, diversity and competitiveness of African countries through a combination of financial and technological support for the development of key infrastructure, access to technology, networking of research centres, adding value to export products, skills development and enhancing market access in support of sustainable development;

(g) Enhance the contribution of the industrial sector, in particular mining, minerals and metals, to the sustainable development of Africa by supporting the development of effective and transparent regulatory and management frameworks and value addition, broad-based participation, social and environmental responsibility and increased market access in order to create an attractive and conducive environment for investment;

(h) Provide financial and technical support to strengthen the capacity of African countries to undertake environmental legislative policy and institutional reform for sustainable development and to undertake environmental impact assessments and, as appropriate, to negotiate and implement multilateral environment agreements;

(i) Develop projects, programmes and partnerships with relevant stakeholders and mobilize resources for the effective implementation of the outcome of the African Process for the Protection and Development of the Marine and Coastal Environment;

(j) Deal effectively with energy problems in Africa, including through initiatives to:

   (i) Establish and promote programmes, partnerships and initiatives to support Africa's efforts to implement NEPAD objectives on energy, which seek to secure access for at least 35 per cent of the African population within 20 years, especially in rural areas;

   (ii) Provide support to implement other initiatives on energy, including the promotion of cleaner and more efficient use of natural gas and increased use of renewable energy, and to improve energy efficiency and access to advanced energy technologies, including cleaner fossil fuel technologies, particularly in rural and peri-urban areas;

(k) Assist African countries in mobilizing adequate resources for their adaptation needs relating to the adverse effects of climate change, extreme weather events, sea level rise and climate variability, and assist in developing national climate change strategies and mitigation programmes, and continue to take actions to mitigate the adverse effects on climate change in Africa, consistent with the United Nations Framework Convention on Climate Change;

(l) Support African efforts to develop affordable transport systems and infrastructure that promote sustainable development and connectivity in Africa;

(m) Further to paragraph 40 above, address the poverty affecting mountain communities in Africa;

(n) Provide financial and technical support for afforestation and reforestation in Africa and to build capacity for sustainable forest management, including combating deforestation and measures to improve the policy and legal framework of the forest sector.

57. Provide financial and technical support for Africa's efforts to implement the Convention to Combat Desertification at the national level and integrate indigenous knowledge systems into land and natural resources management practices, as appropriate, and improve extension services to rural communities and promote better land and watershed management practices, including through improved agricultural practices that address land degradation, in order to develop capacity for the implementation of national programmes.

58. Mobilize financial and other support to develop and strengthen health systems that aim at:

(a) Promoting equitable access to health-care services;

(b) Making available necessary drugs and technology in a sustainable and affordable manner to fight and control communicable diseases, including HIV/AIDS, malaria and tuberculosis, and trypanosomiasis, as well as non-communicable diseases, including those caused by poverty;

(c) Building capacity of medical and paramedical personnel;

(d) Promoting indigenous medical knowledge, as appropriate, including traditional medicine;

(e) Researching and controlling the Ebola disease.

59. Deal effectively with natural disasters and conflicts, including their humanitarian and environmental impacts, recognizing that conflicts in Africa have hindered and in many cases obliterated both the gains and efforts aimed at sustainable development, with the most vulnerable members of society, particularly women and children, being the most impacted victims, through efforts and initiatives, at all levels, to:

(a) Provide financial and technical assistance to strengthen the capacities of African countries, including institutional and human capacity, including at the local level, for effective disaster management, including observation and early warning systems, assessments, prevention, preparedness, response and recovery;

(b) Provide support to African countries to enable them to better deal with the displacement of people as a result of natural disasters and conflicts, and put in place rapid response mechanisms;

(c) Support Africa's efforts for the prevention and resolution, management and mitigation of conflicts and its early response to emerging conflict situations to avert tragic humanitarian consequences;

(d) Provide support to refugee host countries in rehabilitating infrastructure and environment, including ecosystems and habitats, that were damaged in the process of receiving and settling refugees.

60. Promote integrated water resources development and optimize the upstream and downstream benefits therefrom, the development and effective management of water resources across all uses and the protection of water quality and aquatic ecosystems, including through initiatives at all levels, to:

(a) Provide access to potable domestic water, hygiene education and improved sanitation and waste management at the household level through initiatives to encourage public and private investment in water supply and sanitation that give priority to the needs of the poor, within stable and transparent national regulatory frameworks provided by Governments, while respecting local conditions involving all concerned stakeholders and monitoring the performance and improving the accountability of public institutions and private companies; and develop critical water supply, reticulation and treatment infrastructure, and build capacity to maintain and manage systems to deliver water and sanitation services, in both rural and urban areas;

(b) Develop and implement integrated river basin and watershed management strategies and plans for all major water bodies, consistent with paragraph 25 above;

(c) Strengthen regional, subregional and national capacities for data collection and processing, and for planning, research, monitoring, assessment and enforcement, as well as arrangements for water resource management;

(d) Protect water resources, including groundwater and wetland ecosystems, against pollution, as well as, in cases of most acute water scarcity, support efforts for developing nonconventional water resources, including the energy-efficient, cost-effective and sustainable desalination of seawater, rainwater harvesting and recycling of water.

61. Achieve significantly improved sustainable agricultural productivity and food security in furtherance of the agreed millennium development goals, including those contained in the Millennium Declaration, in particular to halve by 2015 the proportion of people who suffer from hunger, including through initiatives at all levels to:

(a) Support the development and implementation of national policies and programmes, including research programmes and development plans of African countries to regenerate their agricultural sector and sustainably develop their fisheries, and increase investment in infrastructure, technology and extension services, according to country needs. African countries should be in the process of developing and implementing food security strategies, within the context of national poverty eradication programmes, by 2005;

(b) Promote and support efforts and initiatives to secure equitable access to land tenure and clarify resource rights and responsibilities, through land and tenure reform processes which respect the rule of law and are enshrined in national law, and to provide access to credit to all, especially to women, and that enable economic and social empowerment and poverty eradication as well as efficient and ecologically sound utilization of land, and enable women producers to become decision makers and owners in the sector, including the right to inherit land;

(c) Improve market access for goods, including goods originating from African countries, in particular least developed countries, within the framework of the Doha Ministerial Declaration, without prejudging the outcome of the WTO negotiations and also within the framework of preferential agreements:

(d) Provide support for African countries to improve regional trade and economic integration between African countries. Attract and increase investment in regional market infrastructure;

(e) Support livestock development programmes aimed at progressive and effective control of animal diseases.

62. Achieve sound management of chemicals, with particular focus on hazardous chemicals and wastes, inter alia, through initiatives to assist African countries in elaborating national chemical profiles, and regional and national frameworks and strategies for chemical management and establishing chemical focal points.

63. Bridge the digital divide and create digital opportunity in terms of access infrastructure and technology transfer and application, through integrated initiatives for Africa. Create an enabling environment to attract investments, accelerate existing and new programmes and projects to connect essential institutions, and stimulate the adoption of information communication technologies in government and commerce programmes and other aspects of national economic and social life.

64. Support Africa's efforts to attain sustainable tourism that contributes to social, economic and infrastructure development through the following measures:

(a) Implementing projects at the local, national and subregional levels, with specific emphasis on marketing African tourism products, such as adventure tourism, eco-tourism and cultural tourism;

(b) Establishing and supporting national and cross-border conservation areas to promote ecosystem conservation according to the ecosystem approach, and to promote sustainable tourism;

(c) Respecting local traditions and cultures and promoting the use of indigenous knowledge in natural resource management and eco-tourism;

(d) Assisting host communities in managing their tourism projects for maximum benefit, while limiting negative impact on their traditions, culture and environment;

(e) Support the conservation of Africa's biological diversity, the sustainable use of its components and the fair and equitable sharing of the benefits arising out of the utilization of genetic resources, in accordance with commitments that countries have under biodiversityrelated a greements to which they are parties, including such agreements as the Convention on Biological Diversity and the Convention on International Trade in Endangered Species of Wild Fauna and Flora, as well as regional biodiversity agreements.

65. Support African countries in their efforts to implement the Habitat Agenda and the Istanbul Declaration through initiatives to strengthen national and local institutional capacities in the areas of sustainable urbanization and human settlements, provide support for adequate shelter and basic services and the development of efficient and effective governance systems in cities and other human settlements, and strengthen, inter alia, the United Nations Human Settlements Programme/UNEP managing water for African cities programme.

OTHER REGIONAL INITIATIVES

66. Important initiatives have been developed within other United Nations regions and regional, subregional and transregional forums to promote sustainable development. The international community welcomes these efforts and the results already achieved, and calls for actions at all levels for their further development, while encouraging interregional, intraregional and international cooperation in this respect, and expresses its support for their further development and implementation by the countries of the regions.

Sustainable development in Latin America and the Caribbean

67. The Initiative of Latin America and the Caribbean on Sustainable Development is an undertaking by the leaders of that region that, building on the Platform for Action on the Road to Johannesburg 2002, which was approved in Rio de Janeiro in October 2001, recognizes the importance of regional actions towards sustainable development and takes into account the region's singularities, shared visions and cultural diversity. It is targeted towards the adoption of concrete actions in different areas of sustainable development, such as biodiversity, water resources, vulnerabilities and sustainable cities, social aspects (including health and poverty), economic aspects

(including energy) and institutional arrangements (including capacitybuilding, indicators and participation of civil society), taking into account ethics for sustainable development.

68. The Initiative envisages the development of actions among countries in the region that may foster South-South cooperation and may count with the support of groups of countries, as well as multilateral and regional organizations, including financial institutions. Being a framework for cooperation, the Initiative is open to partnerships with governments and all major groups.

Sustainable development in Asia and the Pacific

69. Bearing in mind the target of halving the number of people who live in poverty by the year 2015, as provided in the Millennium Declaration, the Phnom Penh Regional Platform on Sustainable Development for Asia and the Pacific recognized that the region contains over half of the world's population and the largest number of the world's people living in poverty. Hence, sustainable development in the region is critical to achieving sustainable development at the global level.

70. The Regional Platform identified seven initiatives for follow-up action: capacity-building for sustainable development; poverty reduction for sustainable development; cleaner production and sustainable energy; land management and biodiversity conservation; protection and management of and access to freshwater resources; oceans, coastal and marine resources and sustainable development of small island developing States; and action on atmosphere and climate change. Follow-up actions of these initiatives will be taken through national strategies and relevant regional and subregional initiatives, such as the Regional Action Programme for Environmentally Sound and Sustainable Development and the Kitakyushu Initiative for a Clean Environment, adopted at the Fourth Ministerial Conference on Environment and Development in Asia and the Pacific organized by the Economic and Social Commission for Asia and the Pacific.

Sustainable development in the West Asia region

71. The West Asia region is known for its scarce water and limited fertile land resources. The region has made progress to a more

knowledge-based production of higher value-added commodities.

72. The regional preparatory meeting endorsed the following priorities: poverty alleviation, relief of debt burden; and sustainable management of natural resources, including, inter alia, integrated water resources management, implementation of programmes to combat desertification, integrated coastal zone management, and land and water pollution control. Sustainable development in the Economic Commission for Europe region

Sustainable Development in the Economic Commission for Europe Region

73. The Economic Commission for Europe (ECE) regional ministerial meeting for the World Summit on Sustainable Development recognized that the region has a major role to play and responsibilities in global efforts to achieve sustainable development by concrete actions. The region recognized that different levels of economic development in countries of the region may require the application of different approaches and mechanisms to implement Agenda-21. In order to address the three pillars of sustainable development in a mutually reinforcing way, the region identified its priority actions for the ECE region for sustainable development in paragraphs 32-46 of a ministerial statement.

74. In furtherance of the region's commitment to sustainable development, there are ongoing efforts at the regional, subregional and transregional levels, including, inter alia, the Environment for Europe process; the fifth ECE ministerial conference, to be held in Kiev in May 2003; the development of an environmental strategy for the 12 countries of Eastern Europe; the Caucasus and Central Asia; the Central Asian Agenda-21; OECD work on sustainable development, the EU sustainable development strategy; and regional and subregional conventions and processes relevant to sustainable development, including, inter alia, the Aarhus Convention, the Alpine Convention, the North American Commission on Environmental Cooperation, the Boundary Waters Treaty, the Iqaluit Declaration of the Arctic Council, the Baltic Agenda-21 and the Mediterranean Agenda-21.

MEANS OF IMPLEMENTATION*

75. The implementation of Agenda-21 and the achievement of the internationally agreed development goals, including those contained in the Millennium Declaration as well as in the present plan of action, require a substantially increased effort, both by countries themselves and by the rest of the international community, based on the recognition that each country has primary responsibility for its own development and that the role of national policies and development strategies cannot be overemphasized, taking fully into account the Rio principles, including, in particular, the principle of common but differentiated responsibilities, which states: "States shall cooperate in a spirit of global partnership to conserve, protect and restore the health and integrity of the Earth's ecosystem. In view of the different contributions to global environmental degradation, States have common but differentiated responsibilities. The developed countries acknowledge the responsibility that they bear in the international pursuit of sustainable development in view of the pressures their societies place on the global environment and of the technologies and financial resources they command." The internationally agreed development goals, including those contained in the Millennium Declaration and Agenda-21, as well as in the present plan of action, will require significant increases in the flow of financial resources as elaborated in the Monterrey Consensus, including through new and additional financial resources, in particular to developing countries, to support the implementation of national policies and programmes developed by them, improved trade opportunities, access to and transfer of environmentally sound technologies on a concessional or preferential basis, as mutually agreed, education and awareness-raising, capacity-building, and information for decision-making and scientific capabilities within the agreed time frame required to meet these goals and initiatives. Progress to this end will require that the international community implement the outcomes of major United Nations

* Because the structure of chapter IX was changed in the course of negotiations, so that its paragraph numbering no longer corresponds to the earlier version contained in document A/CONF.199/L.1, Latin ordinal numbers (bis, ter etc.) are used after paragraph 119 until the end of the chapter in order not to disrupt the paragraph numbering of the following chapter, which was not so changed

conferences, such as the programmes of action adopted at the Third United Nations Conference on the Least Developed Countries, and the Global Conference on the Sustainable Development of Small Island Developing States, and relevant international agreements since 1992, particularly those of the International Conference on Financing for Development and the Fourth WTO Ministerial Conference, including building on them as part of a process of achieving sustainable development.

76. Mobilizing and increasing the effective use of financial resources and achieving the national and international economic conditions needed to fulfil internationally agreed development goals, including those contained in the Millennium Declaration, to eliminate poverty, improve social conditions and raise living standards and protect our environment, will be our first step to ensuring that the twenty-first century becomes the century of sustainable development for all.

77. In our common pursuit of growth, poverty eradication and sustainable development, a critical challenge is to ensure the necessary internal conditions for mobilizing domestic savings, both public and private, sustaining adequate levels of productive investment and increasing human capacity. A crucial task is to enhance the efficacy, coherence and consistency of macroeconomic policies. An enabling domestic environment is vital for mobilizing domestic resources, increasing productivity, reducing capital flight, encouraging the private sector, and attracting and making effective use of international investment and assistance. Efforts to create such an environment should be supported by the international community.

78. Facilitate greater flows of foreign direct investment so as to support the sustainable development activities, including the development of infrastructure, of developing countries, and enhance the benefits that developing countries can draw from foreign direct investment, with particular actions to:

(a) Create the necessary domestic and international conditions to facilitate significant increases in the flow of FDI to developing countries, in particular the least developed countries, which is critical to sustainable development, particularly FDI flows for infrastructure development and other priority areas in developing countries to supplement the domestic resources mobilized by them;

(b) Encourage foreign direct investment in developing countries and countries with economies in transition through export credits that could be instrumental to sustainable development;

79. Recognize that a substantial increase in ODA and other resources will be required if developing countries are to achieve the internationally agreed development goals and objectives, including those contained in the Millennium Declaration. To build support for ODA, we will cooperate to further improve policies and development strategies, both nationally and internationally, to enhance aid effectiveness, with actions to:

(a) Make available the increased ODA commitments announced by several developed countries at the International Conference on Financing for Development. Urge the developed countries that have not done so to make concrete efforts towards the target of 0.7 per cent of GNP as ODA to developing countries, and effectively implement their commitment on ODA to the least developed countries as contained in paragraph 83 of the Programme of Action for the Least Developed Countries for the Decade 2001-2010. We also encourage developing countries to build on progress achieved in ensuring that ODA is used effectively to help achieve development goals and targets in accordance with the outcome of the International Conference on Financing for Development. We acknowledge the efforts of all donors, commend those donors whose ODA contributions exceed, reach or are increasing towards the targets, and underline the importance of undertaking to examine the means and time frames for achieving the targets and goals;

(b) Encourage recipient and donor countries, as well as international institutions, to make ODA more efficient and effective for poverty eradication, sustained economic growth and sustainable development. In this regard, intensify efforts by the multilateral and bilateral financial and development institutions, in accordance with paragraph 43 of the Monterrey Consensus, in particular to harmonize their operational procedures at the highest standards, so as to reduce transaction costs and make ODA disbursement and delivery more flexible and more responsive to the needs of developing countries, taking into account national development needs and objectives under the ownership of recipient countries, and to use development frameworks that are owned and driven by developing countries and that embody poverty reduction strategies, including poverty reduction strategy papers, as vehicles for aid delivery, upon request.

80. Make full and effective use of existing financial mechanisms and institutions, including through actions at all levels to:

(a) Strengthen ongoing efforts to reform the existing international financial architecture, to foster a transparent, equitable and inclusive system that is able to provide for the effective participation of developing countries in the international economic decision-making processes and institutions, as well as for their effective and equitable participation in the formulation of financial standards and codes;
(b) Promote, inter alia, measures in source and destination countries to improve transparency and information about financial flows to contribute to stability in the international financial environment. Measures that mitigate the impact of excessive volatility of short-term capital flows are important and must be considered;
(c) Work to ensure that the funds are made available on a timely, more assured and predictable basis to international organizations and agencies, where appropriate, for their sustainable development activities, programmes and projects;
(d) Encourage the private sector, including transnational corporations, private foundations and civil society institutions, to provide financial and technical assistance to developing countries;
(e) Support new and existing public/private sector financing mechanisms for developing countries and countries with economies in transition, to benefit in particular small entrepreneurs and small, medium-sized and community-based enterprises and to improve their infrastructure, while ensuring the transparency and accountability of such mechanisms.

81. Welcome the successful and substantial third replenishment of the GEF, which will enable it to address the funding requirements of new focal areas and existing ones and continue to be responsive to

the needs and concerns of its recipient countries, in particular developing countries, and further encourage GEF to leverage additional funds from key public and private organizations, improve the management of funds through more speedy and streamlined procedures and simplify its project cycle.

82. Explore ways of generating new public and private innovative sources of finance for development purposes, provided that those sources do not unduly burden developing countries, noting the proposal to use special drawing rights allocations for development purposes, as set forth in paragraph 44 of the Monterrey Consensus.

83. Reduce unsustainable debt burden through such actions as debt relief and, as appropriate, debt cancellation and other innovative mechanisms geared to comprehensively address the debt problems of developing countries, in particular the poorest and most heavily indebted ones. Therefore, debt relief measures should, where appropriate, be pursued vigorously and expeditiously, including within the Paris and London Clubs and other relevant forums, in order to contribute to debt sustainability and facilitate sustainable development, while recognizing that debtors and creditors must share responsibility for preventing and resolving unsustainable debt situations, and that external debt relief can play a key role in liberating resources that can then be directed towards activities consistent with attaining sustainable growth and development. Therefore, we support paragraphs 47 through 51 of the Monterrey Consensus dealing with external debt. Debt relief arrangements should seek to avoid imposing any unfair burdens on other developing countries. There should be an increase in the use of grants for the poorest, debt-vulnerable countries. Countries are encouraged to develop national comprehensive strategies to monitor and manage external liabilities as a key element in reducing national vulnerabilities. In this regard, actions are required to:

(a) Implement speedily, effectively and fully the enhanced heavily indebted poor countries (HIPC) initiative, which should be fully financed through additional resources, taking into consideration, as appropriate, measures to address any fundamental changes in the economic circumstances of those developing countries with unsustainable debt burden caused by natural catastrophes, severe terms-of-trade shocks or affected by conflict, taking into account initiatives which have been undertaken to reduce outstanding indebtedness;

(b) Encourage participation in the HIPC initiative of all creditors that have not yet done so;
(c) Bring international debtors and creditors together in relevant international forums to restructure unsustainable debt in a timely and efficient manner, taking into account the need to involve the private sector in the resolution of crises due to indebtedness, where appropriate;
(d) Acknowledge the problems of the debt sustainability of some non-HIPC low-income countries, in particular those facing exceptional circumstances;
(e) Encourage exploring innovative mechanisms to comprehensively address the debt problems of developing countries, including middle-income countries and countries with economies in transition. Such mechanisms may include debt-for-sustainable-development swaps;
(f) Encourage donor countries to take steps to ensure that resources provided for debt relief do not detract from ODA resources intended to be available for developing countries.

84. Recognizing the major role that trade can play in achieving sustainable development and in eradicating poverty, we encourage WTO members to pursue the work programme agreed at the Fourth WTO Ministerial Conference. In order for developing countries, especially the least developed among them, to secure their share in the growth of world trade commensurate with the needs of their economic development, we urge WTO members to take the following actions:

(a) Facilitate the accession of all developing countries, particularly the least developed countries, as well as countries with economies in transition, that apply for membership of WTO, in accordance with the Monterrey Consensus;
(b) Support the Doha work programme as an important commitment on the part of developed and developing countries to mainstream appropriate trade policies in their respective development policies and programmes;
(c) Implement substantial trade-related technical assistance and capacity-building measures and support the Doha Development Agenda Global Trust Fund established after the Fourth WTO Ministerial Conference as an important step forward in ensuring a sound and predictable basis for WTO-related

technical assistance and capacity-building;

(d) Implement the New Strategy for WTO Technical Cooperation for Capacity-Building, Growth and Integration;

(e) Fully support the implementation of the Integrated Framework for Trade-Related Technical Assistance to Least Developed Countries, and urge development partners to significantly increase contributions to the Trust Fund of the Framework, in accordance with the Doha Ministerial Declaration.

85. In accordance with the Doha Declaration as well as with relevant decisions taken at Doha, we are determined to take concrete action to address issues and concerns raised by developing countries regarding the implementation of some WTO agreements and decisions, including the difficulties and resource constraints faced by them in fulfilling those agreements.

86. Call upon WTO members to fulfil the commitments made in the Doha Ministerial Declaration, notably in terms of market access, in particular for products of export interest to developing countries, especially least developed countries, by implementing the following actions, taking into account paragraph 45 of the Doha Ministerial Declaration:

(a) Review all special and differential treatment provisions with a view to strengthening them and making them more precise, effective and operational, in accordance with paragraph 44 of the Doha Ministerial Declaration;

(b) Aim to reduce or, as appropriate, eliminate tariffs on non-agricultural products, including the reduction or elimination of tariff peaks, high tariffs and tariff escalation, as well as non-tariff barriers, in particular on products of export interest to developing countries. Product coverage should be comprehensive and without a priori exclusions. The negotiations shall take fully into account the special needs and interests of developing and least developed countries, including through less than full reciprocity in reduction commitments, in accordance with the Doha Ministerial Declaration;

(c) Fulfil, without prejudging the outcome of the negotiations, the commitment for comprehensive negotiations initiated under article 20 of the Agreement on Agriculture as referred to in paragraphs 13 and 14 of the Doha Ministerial Declaration, aiming at substantial improvements in market access, reductions of with a view to phasing out all forms of export subsidies, and substantial reductions in trade-distorting domestic support, while agreeing that the provisions for special and differential treatment for developing countries shall be an integral part of all elements of the negotiations and shall be embodied in the schedules of concession and commitments and, as appropriate, in the rules and disciplines to be negotiated, so as to be operationally effective and to enable developing countries to effectively take account of their development needs, including food security and rural development. Take note of the non-trade concerns reflected in the negotiating proposals submitted by WTO members and confirm that non-trade concerns will be taken into account in the negotiations as provided for in the Agreement on Agriculture, in accordance with the Doha Ministerial Declaration.

87. Call on developed countries that have not already done so to work towards the objective of duty-free and quota-free access for all least developed countries' exports, as envisaged in the Programme of Action for the Least Developed Countries for the Decade 2001-2010, which was adopted in Brussels on 20 May 2001.

88. Commit to actively pursue the WTO work programme to address the trade-related issues and concerns affecting the fuller integration of small, vulnerable economies into the multilateral trading system in a manner commensurate with their special circumstances and in support of their efforts towards sustainable development, in accordance with paragraph 35 of the Doha Declaration.

89. Build the capacity of commodity-dependent countries to diversify exports through, inter alia, financial and technical assistance, international assistance for economic diversification and sustainable resource management, and address the instability of commodity prices and declining terms of trade, as well as strengthen the activities covered by the Second Account of the Common Fund for Commodities to support sustainable development.

90. Enhance the benefits for developing countries, as well as countries with economies in transition, from trade liberalization, including through public-private partnerships, through, inter alia, action

at all levels, including through financial support for technical assistance, the development of technology and capacity-building to developing countries to:

(a) Enhance trade infrastructure and strengthen institutions;
(b) Increase developing country capacity to diversify and increase exports to cope with the instability of commodity prices and declining terms of trade;
(c) Increase the value added of developing country exports.

91. Continue to enhance the mutual supportiveness of trade, environment and development with a view to achieving sustainable development through actions at all levels to:

(a) Encourage the WTO Committee on Trade and Environment and the WTO Committee on Trade and Development, within their respective mandates, to each act as a forum to identify and debate developmental and environmental aspects of the negotiations, in order to help achieve an outcome which benefits sustainable development in accordance with the commitments made under the Doha Ministerial Declaration;
(b) Support the completion of the work programme of the Doha Ministerial Declaration on subsidies so as to promote sustainable development and enhance the environment, and encourage reform of subsidies that have considerable negative effects on the environment and are incompatible with sustainable development;
(c) Encourage efforts to promote cooperation on trade, environment and development, including in the field of providing technical assistance to developing countries, between the secretariats of WTO, UNCTAD, UNDP, UNEP and other relevant international environmental and development and regional organizations;

92. In accordance with the Declaration on Principles of International Law concerning Friendly Relations and Cooperation among States in accordance with the Charter of the United Nations, this shall not be construed as authorizing or encouraging any action which would dismember or impair, totally or in part, the territorial integrity or political unity of sovereign and independent States

conducting themselves in compliance with the principle of equal rights and selfdetermination of peoples and thus possessed of a Government representing the whole people belonging to the territory without distinction of any kind.

92.bis. Further resolve to take concerted action against international terrorism, which causes serious obstacles to sustainable development.

* * *

93. Promote, facilitate and finance, as appropriate, access to and the development, transfer and diffusion of environmentally sound technologies and corresponding know-how, in particular to developing countries and countries with economies in transition on favourable terms, including on concessional and preferential terms, as mutually agreed, as set out in chapter 34 of Agenda-21, including through urgent actions at all levels to:

(a) Provide information more effectively;
(b) Enhance existing national institutional capacity in developing countries to improve access to and the development, transfer and diffusion of environmentally sound technologies and corresponding know-how;
(c) Facilitate country-driven technology needs assessments;
(d) Establish legal and regulatory frameworks in both supplier and recipient countries that expedite the transfer of environmentally sound technologies in a cost-effective manner by both public and private sectors and support their implementation;
(e) Promote the access and transfer of technology related to early warning systems and to mitigation programmes to developing countries affected by natural disasters.

94. Improve the transfer of technologies to developing countries, in particular at the bilateral and regional levels, including through urgent actions at all levels to:

(a) Improve interaction and collaboration, stakeholder relationships and networks between and among universities, research institutions, government agencies and the private sector;

(b) Develop and strengthen networking of related institutional support structures, such as technology and productivity centres, research, training and development institutions, and national and regional cleaner production centres;

(c) Create partnerships conducive to investment and technology transfer, development and diffusion, to assist developing countries, as well as countries with economies in transition, in sharing best practices and promoting programmes of assistance, and encourage collaboration between corporations and research institutes to enhance industrial efficiency, agricultural productivity, environmental management and competitiveness;

(d) Provide assistance to developing countries, as well as countries with economies in transition, in accessing environmentally sound technologies that are publicly owned or in the public domain, as well as available knowledge in the public domain on science and technology, and in accessing the know-how and expertise required in order for them to make independent use of this knowledge in pursuing their development goals;

(e) Support existing mechanisms and, where appropriate, establish new mechanisms for the development, transfer and diffusion of environmentally sound technologies to developing countries and economies in transition.

* * *

95. Assist developing countries in building capacity to access a larger share of multilateral and global research and development programmes. In this regard, strengthen and, where appropriate, create centres for sustainable development in developing countries.

96. Build greater capacity in science and technology for sustainable development, with action to improve collaboration and partnerships on research and development and their widespread application among research institutions, universities, the private sector, governments, NGOs and networks, as well as between and among scientists and academics of developing and developed countries, and in this regard encourage networking with and between centres of scientific excellence in developing countries.

97. Improve policy and decision-making at all levels through, inter alia, improved collaboration between natural and social scientists, and between scientists and policy makers, including through urgent actions

at all levels to:

(a) Increase the use of scientific knowledge and technology, and increase the beneficial use of local and indigenous knowledge in a manner respectful of the holders of that knowledge and consistent with national law;
(b) Make greater use of integrated scientific assessments, risk assessments and interdisciplinary and intersectoral approaches;
(c) Continue to support and collaborate with international scientific assessments supporting decision-making, including the Intergovernmental Panel on Climate Change, with the broad participation of developing country experts;
(d) Assist developing countries in developing and implementing science and technology policies;
(e) Establish partnerships between scientific, public and private institutions, and by integrating scientists' advice into decision-making bodies in order to ensure a greater role for science, technology development and engineering sectors;
(f) Promote and improve science-based decision-making and reaffirm the precautionary approach as set out in principle 15 of the Rio Declaration on Environment and Development, which states: "In order to protect the environment, the precautionary approach shall be widely applied by States according to their capabilities. Where there are threats of serious or irreversible damage, lack of full scientific certainty shall not be used as a reason for postponing cost-effective measures to prevent environmental degradation."

98. Assist developing countries, through international cooperation, in enhancing their capacity in their efforts to address issues pertaining to environmental protection including in their formulation and implementation of policies for environmental management and protection, including through urgent actions at all levels to:

(a) Improve their use of science and technology for environmental monitoring, assessment models, accurate databases and integrated information systems;
(b) Promote and, where appropriate, improve their use of satellite technologies for quality data collection, verification and

updating, and further improvement of aerial and groundbased observations, in support of their efforts to collect quality, accurate, long-term, consistent and reliable data;

99. Eliminate gender disparity in primary and secondary education by 2005, as provided in the Dakar Framework for Action on Education for All, and at all levels of education no later than 2015, to meet the development goals contained in the Millennium Declaration, with action to ensure, inter alia, equal access to all levels and forms of education, training and capacitybuilding by gender mainstreaming, and by creating a gender-sensitive educational system.

100. Integrate sustainable development into education systems at all levels of education in order to promote education as a key agent for change.

101. Develop, implement, monitor and review education action plans and programmes at the national, subnational and local levels, as appropriate, that reflect the Dakar Framework for Action on Education for All and that are relevant to local conditions and needs leading to the achievement of community development, and make education for sustainable development a part of those plans.

102. Provide all community members with a wide range of formal and non-formal continuing educational opportunities, including volunteer community service programmes, in order to end illiteracy and emphasize the importance of lifelong learning and promote sustainable development.

103. Support the use of education to promote sustainable development, including through urgent actions at all levels to:

(a) Integrate information and communications technology in school curriculum development to ensure its access by both rural and urban communities, and provide assistance particularly to developing countries, inter alia, for the establishment of an appropriate enabling environment required for such technology;
(b) Promote, as appropriate, affordable and increased access to programmes for students, researchers and engineers from developing countries in the universities and research institutions of developed countries in order to promote the exchange of experience and capacity that will benefit all partners;

(c) Continue to implement the work programme of the Commission on Sustainable Development on education for sustainable development;
(d) Recommend to the United Nations General Assembly that it consider adopting a decade of education for sustainable development, starting in 2005.

* * *

104. Enhance and accelerate human, institutional and infrastructure capacity-building initiatives, and promote partnerships in that regard that respond to the specific needs of developing countries in the context of sustainable development.

105. Support local, national, subregional and regional initiatives, with action to develop, use and adapt knowledge and techniques and to enhance local, national, subregional and regional centres of excellence for education, research and training in order to strengthen the knowledge capacity of developing countries and countries with economies in transition through, inter alia, the mobilization from all sources of adequate financial and other resources, including new and additional resources;

105.bis Provide technical and financial assistance to developing countries, including through the strengthening of capacity-building efforts, such as the United Nations Development Programme capacity 21 programme, to:

(a) Assess their own capacity development needs and opportunities at the individual, institutional and societal levels;
(b) Design programmes for capacity-building and support for local, national and community-level programmes that focus on meeting the challenges of globalization more effectively and attaining the internationally agreed development goals, including those contained in the Millennium Declaration;
(c) Develop the capacity of civil society, including youth, to participate, as appropriate, in designing, implementing and reviewing sustainable development policies and strategies at all levels;
(d) Build and, where appropriate, strengthen national capacities for carrying out effective implementation of Agenda-21.

* * *

105.ter Ensure access, at the national level, to environmental information and judicial and administrative proceedings in environmental matters, as well as public participation in decision making, so as to further principle 10 of the Rio Declaration on Environment and Development, taking into full account principles 5, 7 and 11 of the Declaration.

105.quater Strengthen national and regional information and statistical and analytical services relevant to sustainable development policies and programmes, including data disaggregated by sex, age and other factors, and encourage donors to provide financial and technical support to developing countries to enhance their capacity to formulate policies and implement programmes for sustainable development.

105.quinquies Encourage further work on indicators for sustainable development by countries at the national level, including integration of gender aspects, on a voluntary basis, in line with national conditions and priorities.

105.sexties Promote further work on indicators, in conformity with paragraph 3 of decision 9/4 of the Commission on Sustainable Development.

105.septies Promote the development and wider use of earth observation technologies, including satellite remote sensing, global mapping and geographic information systems, to collect quality data on environmental impacts, land use and land-use changes, including through urgent actions at all levels to:

(a) Strengthen cooperation and coordination among global observing systems and research programmes for integrated global observations, taking into account the need for building capacity and sharing of data from ground-based observations, satellite remote sensing and other sources among all countries;
(b) Develop information systems that make the sharing of valuable data possible, including the active exchange of Earth observation data;
(c) Encourage initiatives and partnerships for global mapping.

105.octies Support countries, particularly developing countries, in their national efforts to:

(a) Collect data that are accurate, long-term, consistent and reliable;

(b) Use satellite and remote-sensing technologies for data collection and further improvement of ground-based observations;

(c) Access, explore and use geographic information by utilizing the technologies of satellite remote sensing, satellite global positioning, mapping and geographic information systems.

105.noviens Support efforts to prevent and mitigate the impacts of natural disasters, including through urgent actions at all levels to:

(a) Provide affordable access to disaster-related information for early warning purposes;

(b) Translate available data, particularly from global meteorological observation systems, into timely and useful products.

105.diciens Develop and promote the wider application of environmental impact assessments, inter alia, as a national instrument, as appropriate, to provide essential decision-support information on projects that could cause significant adverse effects to the environment.

105.undeciens Promote and further develop methodologies at policy, strategy and project levels for sustainable development decision-making at the local and national levels, and where relevant at the regional level. In this regard, emphasize that the choice of the appropriate methodology to be used in countries should be adequate to their country-specific conditions and circumstances, should be on a voluntary basis and should conform to their development priority needs.

INSTITUTIONAL FRAMEWORK FOR SUSTAINABLE DEVELOPMENT

106. An effective institutional framework for sustainable development at all levels is key to the full implementation of Agenda-21, the follow-up to the outcomes of the World Summit on Sustainable Development and meeting emerging sustainable development challenges. Measures aimed at strengthening such a framework should build on the provisions of Agenda-21 as well as the 1997 Programme for its further implementation and the principles of the Rio Declaration on Environment and Development and should promote the achievement of the internationally agreed development goals, including those

contained in the Millennium Declaration, taking into account the Monterrey Consensus and relevant outcomes of other major United Nations conferences and international agreements since 1992. It should be responsive to the needs of all countries, taking into account the specific needs of developing countries including the means of implementation. It should lead to the strengthening of international bodies and organizations dealing with sustainable development, while respecting their existing mandates, as well as to the strengthening of relevant regional, national and local institutions.

106.bis. Good governance is essential for sustainable development. Sound economic policies, solid democratic institutions responsive to the needs of the people and improved infrastructure are the basis for sustained economic growth, poverty eradication, and employment creation. Freedom, peace and security, domestic stability, respect for human rights, including the right to development, and the rule of law, gender equality, market-oriented policies, and an overall commitment to just and democratic societies are also essential and mutually reinforcing.

Objectives

107. Measures to strengthen sustainable development institutional arrangements at all levels should be taken within the framework of Agenda-21* and should build on developments since UNCED, and should lead to the achievement of, inter alia, the following objectives:

(a) Strengthening commitments to sustainable development;
(b) Integration of the economic, social and environmental dimensions of sustainable development in a balanced manner;
(c) Strengthening of the implementation of Agenda-21, including through the mobilization of financial and technological resources, as well as capacity-building programmes, particularly for developing countries;
(d) Strengthening coherence, coordination and monitoring;
(e) Promoting the rule of law and strengthening of governmental institutions;

---

* References in the present chapter to Agenda-21 are deemed to include Agenda-21, the Programme for the Further Implementation of Agenda-21 and the outcomes of the Summit.

(f) Increasing effectiveness and efficiency through limiting overlap and duplication of activities of international organizations, within and outside the United Nations system, based on their mandates and comparative advantages;
(g) Enhancing participation and effective involvement of civil society and other relevant stakeholders in the implementation of Agenda-21, as well as promoting transparency and broad public participation;
(h) Strengthening capacities for sustainable development at all levels, including the local level, in particular those of developing countries;
(i) Strengthening international cooperation aimed at reinforcing the implementation of Agenda-21 and the outcomes of the Summit.

Strengthening the institutional framework for sustainable development at the international level

108. The international community should:

(a) Enhance the integration of sustainable development goals as reflected in Agenda-21 and support for implementation of Agenda-21 and the outcomes of the Summit into the policies, work programmes and operational guidelines of relevant United Nations agencies, programmes and funds, GEF and international financial and trade institutions within their mandates, while stressing that their activities should take full account of national programmes and priorities, particularly those of developing countries, as well as, where appropriate, countries with economies in transition, to achieve sustainable development;
(b) Strengthen collaboration within and between the United Nations system, international financial institutions, the Global Environment Facility and WTO, utilizing the United Nations Chief Executives Board for Coordination (CEB), the United Nations Development Group, the Environment Management Group and other inter-agency coordinating bodies. Strengthened inter-agency collaboration should be pursued in all relevant contexts, with special emphasis on the operational level and involving partnership arrangements on

specific issues to support, in particular, developing countries' efforts in implementing Agenda-21;

(c) Strengthen and better integrate the three dimensions of sustainable development policies and programmes, and promote the full integration of sustainable development objectives into programmes and policies of bodies that have a primary focus on social issues. In particular, the social dimension of sustainable development should be strengthened, inter alia, by emphasizing follow-up to the outcomes of the World Summit for Social Development and its five-year review, and taking into account their reports,.and by support to social protection systems;

(d) Fully implement the outcomes of decision I on international environmental governance adopted by the UNEP Governing Council at its seventh special session, and invite the General Assembly at its fifty-seventh session to consider the important but complex issue of establishing universal membership for the Governing Council/Global Ministerial Environment Forum;

(e) Engage actively and constructively in ensuring the timely completion of the negotiations on a comprehensive United Nations convention against corruption, including the question of repatriation of funds illicitly acquired to countries of origin;

(f) Promote corporate responsibility and accountability and the exchange of best practices in the context of sustainable development, including, as appropriate, through multi-stakeholder dialogue, such as through the Commission on Sustainable Development, and other initiatives;

(g) Take concrete action to implement the Monterrey Consensus at all levels.

109. Good governance at the international level is fundamental for achieving sustainable development. In order to ensure a dynamic and enabling international economic environment, it is important to promote global economic governance through addressing the international finance, trade, technology and investment patterns that have an impact on the development prospects of developing countries. To this effect, the international community should take all necessary and appropriate measures, including ensuring support for structural and macroeconomic reform, a comprehensive solution to the external

debt problem and increasing market access for developing countries. Efforts to reform the international financial architecture need to be sustained with greater transparency and the effective participation of developing countries in decision-making processes. A universal, rule-based, open, non-discriminatory and equitable multilateral trading system, as well as meaningful trade liberalization, can substantially stimulate development worldwide, benefiting countries at all stages of development.

110. A vibrant and effective United Nations system is fundamental to the promotion of international cooperation for sustainable development and to a global economic system that works for all. To this effect, a firm commitment to the ideals of the United Nations and to the principles of international law and those enshrined in the Charter of the United Nations, and to strengthening the United Nations system and other multilateral institutions and promoting the improvement of their operations, is essential. States should also fulfil their commitment to negotiate and finalize as soon as possible a United Nations convention against corruption in all its aspects, including the question of repatriation of funds illicitly acquired to countries of origin and also to promoting stronger cooperation to eliminate money-laundering.

Role of the General Assembly

111. The General Assembly of the United Nations should adopt sustainable development as a key element of the overarching framework for United Nations activities, particularly for achieving the internationally agreed development goals, including those contained in the Millennium Declaration, and should give overall political direction to the implementation of Agenda-21 and its review.

Role of the Economic and Social Council

112. Pursuant to the relevant provisions of the Charter of the United Nations and Agenda-21 provisions regarding the Economic and Social Council and General Assembly resolutions 48/162 and 50/227, which reaffirmed the Council as the central mechanism for the coordination of the United Nations system and its specialized agencies and supervision of subsidiary bodies, in particular its functional commissions, and to promote the implementation of Agenda-21 by strengthening system-wide coordination, the Council should:

(a) Increase its role in overseeing system-wide coordination and the balanced integration of economic, social and environmental aspects of United Nations policies and programmes aimed at promoting sustainable development;

(b) Organize periodic consideration of sustainable development themes in regard to the implementation of Agenda-21, including the means of implementation. Recommendations in regard to such themes could be made by the Commission on Sustainable Development;

(c) Make full use of its high-level, coordination, operational activities and the general segments to effectively take into account all relevant aspects of the work of the United Nations on sustainable development. In this context, the Council should encourage the active participation of major groups in its high-level segment and the work of its relevant functional commissions, in accordance with the respective rules of procedure;

(d) Promote greater coordination, complementarity, effectiveness and efficiency of activities of its functional commissions and other subsidiary bodies that are relevant to the implementation of Agenda-21;

(e) Terminate the work of the Committee on Energy and Natural Resources for Development and transfer its work to the Commission on Sustainable Development;

(f) Ensure that there is a close link between the role of the Council in the follow-up to the Summit and its role in the follow-up to the Monterrey Consensus, in a sustained and coordinated manner. To that end, the Council should explore ways to develop arrangements relating to its meetings with the Bretton Woods institutions and WTO, as set out in the Monterrey Consensus;

(g) Intensify its efforts to ensure that gender mainstreaming is an integral part of its activities concerning the coordinated implementation of Agenda-21.

Role and function of the Commission on Sustainable Development

113. The Commission on Sustainable Development should continue to be the high-level commission on sustainable development within the United Nations system and serve as a forum for consideration of issues related to integration of the three dimensions of sustainable

development. Although the role, functions and mandate of the Commission as set out in relevant parts of Agenda-21 and adopted in General Assembly resolution 47/191 continue to be relevant, the Commission needs to be strengthened, taking into account the role of relevant institutions and organizations. An enhanced role of the Commission should include reviewing and monitoring progress in the implementation of Agenda-21 and fostering coherence of implementation, initiatives and partnerships.

114. Within that context, the Commission should give more emphasis on actions that enable implementation at all levels, including promoting and facilitating partnerships involving Governments, international organizations and relevant stakeholders for the implementation of Agenda-21.

115. The Commission should:

(a) Review and evaluate progress and promote further implementation of Agenda-21;
(b) Focus on the cross-sectoral aspects of specific sectoral issues and provide a forum for better integration of policies, including through interaction among Ministers dealing with the various dimensions and sectors of sustainable development through the high-level segments;
(c) Address new challenges and opportunities related to the implementation of Agenda-21;
(d) The Commission should focus on actions related to implementation of Agenda-21, limiting negotiations in the sessions of the Commission to every two years;
(e) Limit the number of themes addressed in each session.

116. In relation to its role in facilitating implementation, the Commission should emphasize the following:

(a) Review progress and promote the further implementation of Agenda-21. In this context, the Commission should identify constraints on implementation and make recommendations to overcome those constraints;
(b) Serve as a focal point for the discussion of partnerships that promote sustainable development, including sharing lessons learned, progress made and best practices;
(c) Review issues related to financial assistance and transfer of

technology for sustainable development, as well as capacity-building, while making full use of existing information. In this regard, the Commission on Sustainable Development could give consideration to more effective use of national reports and regional experience and to this end make appropriate recommendations;

(d) Provide a forum for analysis and exchange of experience on measures that assist sustainable development planning, decision-making and the implementation of sustainable development strategies. In this regard, the Commission could give consideration to more effective use of national and regional reports;

(e) Take into account significant legal developments in the field of sustainable development, with due regard to the role of relevant intergovernmental bodies in promoting the implementation of Agenda-21 relating to international legal instruments and mechanisms.

117. With regard to the practical modalities and programme of work of the Commission, specific decisions on those issues should be taken by the Commission at its next session, when the Commission's thematic work programme will be elaborated. In particular, the following issues should be considered:

(a) Giving a balanced consideration to implementation of all of the mandates of the Commission contained in General Assembly resolution 47/191;

(b) Continuing to provide for more direct and substantive involvement of international organizations and major groups in the work of the Commission;

(c) Give greater consideration to the scientific contributions to sustainable development through, for example, drawing on the scientific community and encouraging national, regional and international scientific networks to be involved in the Commission;

(d) Furthering the contribution of educators to sustainable development, including, where appropriate, in the activities of the Commission;

(e) The scheduling and duration of intersessional meetings.

118. Undertake further measures to promote best practices and

lessons learned in sustainable development, and in addition promote the use of contemporary methods of data collection and dissemination, including broader use of information technologies.

Role of international institutions

119. Stress the need for international institutions both within and outside the United Nations system, including international financial institutions, WTO and GEF, to enhance, within their mandates, their cooperative efforts to:

(a) Promote effective and collective support to the implementation of Agenda-21 at all levels;
(b) Enhance the effectiveness and coordination of international institutions to implement Agenda-21, the outcomes of the World Summit on Sustainable Development, relevant sustainable development aspects of the Millennium Declaration, the Monterrey Consensus and the outcomes of the fourth WTO ministerial meeting, held in Doha in November 2001.

120. Request the Secretary-General of the United Nations, utilizing the United Nations System Chief Executives Board for Coordination, including through informal collaborative efforts, to further promote system-wide inter-agency cooperation and coordination on sustainable development, to take appropriate measures to facilitate exchange of information, and to continue to keep the Economic and Social Council and the Commission informed of actions being taken to implement Agenda-21.

121. Significantly strengthen support for UNDP capacity-building programmes for sustainable development, building on the experience gained from Capacity 21, as important mechanisms supporting local and national development capacity-building efforts, in particular in developing countries.

122. Strengthen cooperation among UNEP and other United Nations bodies and specialized agencies, the Bretton Woods institutions and WTO, within their mandates.

123. UNEP, UN-Habitat, UNDP and UNCTAD, within their mandates, should strengthen their contribution to sustainable development programmes and the implementation of Agenda-21 at all

levels, particularly in the area of promoting capacity-building.

124. To promote effective implementation of Agenda-21 at the international level, the following should also be undertaken:

(a) Streamline the international sustainable development meeting calendar and, as appropriate, reduce the number of meetings, the length of meetings and the amount of time spent on negotiated outcomes in favour of more time spent on practical matters related to implementation;
(b) Encourage partnership initiatives for implementation by all relevant actors to support the outcome of the World Summit on Sustainable Development. In this context, further development of partnerships and partnership follow-up should take note of the preparatory work for the Summit;
(c) Make full use of developments in the field of information and communication technologies.

125. Strengthening of the international institutional framework for sustainable development is an evolutionary process. It is necessary to keep under review relevant arrangements; identify gaps; eliminate duplication of functions; and continue to strive for greater integration, efficiency and coordination of the economic, social and environmental dimensions of sustainable development aiming at the implementation of Agenda-21.

Strengthening institutional arrangements for sustainable development at the regional level

126. Implementation of Agenda-21 and the outcomes of the Summit should be effectively pursued at the regional and subregional levels, through the regional commissions and other regional and subregional institutions and bodies.

127. Intraregional coordination and cooperation on sustainable development should be improved among the regional commissions, United Nations Funds, programmes and agencies, regional development banks, and other regional and subregional institutions and bodies. This should include, as appropriate, support for development, enhancement and implementation of agreed regional sustainable development strategies and action plans, reflecting national and regional priorities.

128. In particular and taking into account relevant provisions of

Agenda-21, the regional commissions, in collaboration with other regional and subregional bodies, should:

(a) Promote the integration of the three dimensions of sustainable development into their work in a balanced way, including through implementation of Agenda-21. To this end, the regional commissions should enhance their capacity through internal action and be provided, as appropriate, with external support;
(b) Facilitate and promote a balanced integration of the economic, social and environmental dimensions of sustainable development into the work of regional, subregional and other bodies, for example by facilitating and strengthening the exchange of experiences, including national experience, best practices, case studies and partnership experience related to the implementation of Agenda-21;
(c) Assist in the mobilization of technical and financial assistance, and facilitate the provision of adequate financing for the implementation of regionally and subregionally agreed sustainable development programmes and projects, including addressing the objective of poverty eradication;
(d) Continue to promote multi-stakeholder participation and encourage partnerships to support the implementation of Agenda-21 at the regional and subregional levels.

129. Regionally and subregionally agreed sustainable development initiatives and programmes, such as the New Partnership for Africa's Development (NEPAD) and the interregional aspects of the globally agreed Programme of Action for the Sustainable Development of Small Island Developing States, should be supported.

Strengthening institutional frameworks for sustainable development at the national level

130. States should:

(a) Continue to promote coherent and coordinated approaches to institutional frameworks for sustainable development at all national levels, including through, as appropriate, the establishment or strengthening of existing authorities and mechanisms necessary for policymaking, coordination and implementation and enforcement of laws;

(b) Take immediate steps to make progress in the formulation and elaboration of national strategies for sustainable development and begin their implementation by 2005. To this end, as appropriate, strategies should be supported through international cooperation, taking into account the special needs of developing countries, in particular the least developed countries. Such strategies, which, where applicable, could be formulated as poverty reduction strategies that integrate economic, social and environmental aspects of sustainable development, should be pursued in accordance with each country's national priorities.

131. Each country has the primary responsibility for its own sustainable development, and the role of national policies and development strategies cannot be overemphasized. All countries should promote sustainable development at the national level by, inter alia, enacting and enforcing clear and effective laws that support sustainable development. All countries should strengthen governmental institutions, including by providing necessary infrastructure and by promoting transparency, accountability and fair administrative and judicial institutions.

131.bis All countries should also promote public participation, including through measures that provide access to information regarding legislation, regulations, activities, policies and programmes. They should also foster full public participation in sustainable development policy formulation and implementation. Women should be able to participate fully and equally in policy formulation and decision-making.

132. Further promote the establishment or enhancement of sustainable development councils and/or coordination structures at the national level, including at the local level, in order to provide a high-level focus on sustainable development policies. In that context, multistakeholder participation should be promoted.

133. Support efforts by all countries, particularly developing countries, as well as countries with economies in transition, to enhance national institutional arrangements for sustainable development, including at the local level. That could include promoting cross-sectoral approaches in the formulation of strategies and plans for sustainable development, such as, where applicable, poverty reduction strategies,

aid coordination, encouraging participatory approaches and enhancing policy analysis, management capacity and implementation capacity, including mainstreaming a gender perspective in all those activities.

134. Enhance the role and capacity of local authorities as well as stakeholders in implementing Agenda-21 and the outcomes of the Summit and in strengthening the continuing support for local Agenda-21 programmes and associated initiatives and partnerships, and encourage, in particular, partnerships among and between local authorities and other levels of government and stakeholders to advance sustainable development as called for in, inter alia, the Habitat Agenda.

Participation of major groups

135. Enhance partnerships between governmental and non-governmental actors, including all major groups, as well as volunteer groups, on programmes and activities for the achievement of sustainable development at all levels.

136. Acknowledge the consideration being given to the possible relationship between environment and human rights, including the right to development, with full and transparent participation of Member States of the United Nations and observer States.

137. Promote and support youth participation in programmes and activities relating to sustainable development through, for example, supporting local youth councils or their equivalent, and by encouraging their establishment where they do not exist.

INDEX